KEYS
TO
MATHEMATICS

JOHN E. MAXFIELD

Professor of Mathematics
Kansas State University

and

MARGARET W. MAXFIELD

1973

W. B. SAUNDERS COMPANY · PHILADELPHIA · LONDON · TORONTO

W. B. Saunders Company: West Washington Square
Philadelphia, PA 19105

12 Dyott Street
London, WC1A 1DB

833 Oxford Street
Toronto18, Ontario

Keys to Mathematics

ISBN 0-7216-6193-9

Print No: 9   8   7   6   5   4   3   2   1

PREFACE

This book is for all general education students, with special helps for those who have a learning handicap in mathematics:

students who have had no formal mathematics course since arithmetic,

students who have forgotten the mathematics they have studied, perhaps because of an interruption in their schooling,

students whose earlier experience has been unpleasant, leaving them with a negative, fearful attitude,

students who want to learn what mathematics has to offer besides computational skills.

How can a general education mathematics course meet the needs of these students?

The topics should be of real value and importance, central to a modern mathematical viewpoint, but they should be new to the students' experiences with mathematics courses.

The emphasis should not be on diagnosing deficiencies and offering remedial work, but on helping students realize how much mathematics they have already absorbed informally, and how frequently mathematics enters their lives in noncomputational forms.

The presentation should be complete and simplified enough to assure a successful, pleasant experience for students who interact with the course to the extent of working the exercises.

The course should leave students with a legitimate feeling of accomplishment and with confidence in their ability to learn mathematics and put it to use.

How can this book help achieve the goals of a general education course?

The ten topics chosen are new to most students as topics for mathematics. However, as their importance suggests, they arise in fairly common applications. In this treatment, students have the reassurance of known areas of application together with novelty in the approach.

The "Key" statement at the end of each chapter shows how the topic contributes to understanding mathematics as a whole.

Topics are introduced through familiar settings, and the reader reinforces what he has already observed before going on to a more abstract formulation. This applications-first approach facilitates learning and also provides motivation.

Perhaps the most crippling learning problem in mathematics is lack of confidence. Many students, if they can be freed from fear that they will get the wrong answer, can cope well with mathematics problems from their own sense and general knowledge, and in fact do so in everyday life outside the classroom. In this book the student is encouraged to rely on his own judgment and to trust his own observation. Some problems specifically invite the student to discuss his own experiences and opinions. Especially in Chapter 1, a result is not *wrong*, even if it is not a *best* or *optimal* solution, and if it does not satisfy the student, he simply tries again. This experimental, rather than theoretical, attack usually stays with the student and enables him to try new methods.

JOHN E. MAXFIELD

MARGARET W. MAXFIELD

ACKNOWLEDGMENTS

The project of developing mathematics text material for general education students has been so on our minds and lips that few friends have escaped. Some have offered generous advice, even detailed reconstructions, as professional mathematicians and teachers. Some have volunteered as completely naive readers to check the level of exposition. Some have delighted us by mentioning related material from diverse fields.

First, let us thank our reviewers. Joan and James Leitzel of Ohio State University worked through the entire book and suggested alternative presentations for several items, as well as many minor changes.

Professor Richard A. Duke of the University of Washington, Professor Anthony A. Gioia of Western Michigan University, Professor Charles C. Lindner of Auburn University, and Professor William H. Caldwell of the University of South Carolina reviewed the book as a whole, and helped us with perspective and consistency.

Sections of one or two chapters were given detailed reviews by Professors William H. Caldwell, Robert W. Floyd, David S. Moore, Herbert J. Nichol, Peter L. Renz, and Beth Unger.

In addition, many kind friends read parts of the book and made helpful comments: Doris Chappell, Joan and Tim Cowan, Emily Danskin, Nancy Hause (who brought to our attention the Girl Scout process used in Chapter 1), Barbara MacMillan, the Manhattan High School Mathematics Club, and Elaine and Nancy Maxfield. Thanks to Paul Bergstresser, M.D., for letting us explain so much about the college-admissions process that he recognized in it the Intern Matching Program, and told us about it.

David Maxfield took lots of photographs for us, and helped to plan them so as to illustrate exactly the points we had in mind. Lori Alexander, Ralph Currie, Sheri Feltner, Kim Meyer, John Nordin, Joe Poole, Greg Thompson, and Dave Weixelman served as models.

Grant Lashbrook, Art Director at the W. B. Saunders Company, developed cartoons, drawings, and diagrams in close cooperation with the authors. He was able to suggest visual devices for clarifying several points.

All publishers, authors, and organizations who have been

asked for permission to use material have granted that permission readily and many have gone out of their way to send photographs, additional brochures, even books to help us and our readers.

Of course, some of our best helpers have been students at the University of Florida and at Kansas State University, who have taught us what is easy and what is hard, what is necessary and what can be omitted, what is interesting, and what is boring.

Even people not sufficiently knowledgeable about architecture to be critical can be influenced by the shape and arrangement of the buildings they live in. Readers of textbooks, although they may not understand the principles of book design, can receive an important aid in uncluttered, attractive pages. Lorraine Battista, manager of the W. B. Saunders design department, has been alert to every possibility for giving us this extra help.

Thanks to Melvin Black, Pamela Herr, Ray Kersey, Herbert Powell, Jr., and others at Saunders, all of whom seemed to take a special interest in the book.

Thank you, George Fleming, mathematics editor for KEYS. From your initial suggestion that we put it on paper, until the production department took over, you have flattered, harangued, doubted, and reassured, as needed at each time, and have done each superbly.

J. E. M.
M. W. M.

CONTENTS

KEYS
TO
MATHEMATICS

CHAPTER 1

A British lady living in India was disgusted because a workman seemed unable to figure out how to install her telephone. Finally she told him, "There. If you just move that little piece over to here, and then thread that through there, it should work." The solution she described worked perfectly, and she continued, "I should think you could have figured that out from common sense."

"Madam," he replied with dignity, "common sense is a rare gift of God; all I have is a technical education."

Here we explore some stepwise processes for solving problems without depending on quantification and numerical calculations. These processes, some of them quite new, adapt well to social science problems for which quantification is sometimes artificial. The vocabulary and techniques help us describe and analyze interrelationships.

NETWORKS AND GRAPHS

Common-Sense Mathematics

STABLE ASSIGNMENTS

Dating Service

You run a small dating service. Your current problem is to arrange dates for 4 boys with 4 girls. The boys are Al, Bob, Cal, and Don. (You can use code names a, b, c, d.) The girls are Alice, Betty, Cathy, and Diana (A, B, C, D). Each boy has filled out a questionnaire on which he has ranked the 4 girls in order of preference, from rank 1 for his favorite to rank 4 for the girl he likes least. For instance, Al filled out his questionnaire to read

Al: first choice, Betty
second choice, Diana
third choice, Alice
fourth choice, Cathy

His rankings appear in the first horizontal row of Figure 1–1, the first entry in each case.

		Alice	Betty	Cathy	Diana
Al	a	3, __	1, __	4, __	2, __

Each girl has filled out a questionnaire ranking the boys. Cathy's, for instance, reads

Cathy: first choice, Don
second choice, Al
third choice, Bob
fourth choice, Cal

These rankings appear in the third vertical column of Figure 1–1, the second entry in each case.

	Cathy C
Al	—, 2
Bob	—, 3
Cal	—, 4
Don	—, 1

Figure 1–1 sums up all eight questionnaires.

		Alice A	Betty B	Cathy C	Diana D
Al	a	3, 1	1, 2	4, 2	2, 1
Bob	b	2, 3	3, 3	4, 3	1, 3
Cal	c	3, 2	2, 1	4, 4	1, 2
Don	d	3, 4	1, 4	4, 1	2, 4

FIGURE 1–1. RANKINGS OF GIRLS BY BOYS, THEN BOYS BY GIRLS.

Exercise 1–1. Which one of the girls does Bob like best? least? second-best?

Exercise 1–2. Is every boy the favorite of one of the girls? Which girls are favorites for two different boys?

The object of your dating service is to avoid unstable schedules of dates, which you define as follows:

> **Definition 1–1.** A schedule matching each boy with one girl is **unstable** if for some boy Adam and some girl Eve, Adam is not scheduled to date Eve, yet Adam would prefer Eve to his date and Eve would prefer Adam to her date.

First you check whether you can give each boy his first choice. This would call for having Al date Betty, Bob date Diana, Cal date Diana, and Don date Betty. You sketch this on the blackboard by linking appropriately labeled dots, as in Figure 1–2a. You notice right away that Betty

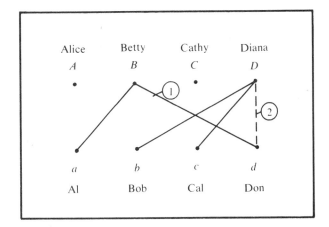

FIGURE 1–2a.

would have 2 dates, Diana would have 2 dates, and Alice and Cathy would have no dates, so that this first trial scheduling fails. Since Betty cannot date both Al and Don, reject the one she ranks lower in preference, which from Figure 1–1 is Don. On the blackboard you erase the link from Don to Betty, as indicated by ① in Figure 1–2a. Then you try giving Don his second choice, which from Figure 1–1 is Diana.

Now work on the conflict over Diana, whom Bob, Cal, and Don want to date. Reject Don for her, since she rated him 4 in preference. Erase the link from Don to Diana.

Then Don must try his next choice after Diana, namely Alice. Figure 1–2b reflects the next few changes in your blackboard sketch.

Exercise 1–3. Explain Figure 1–2b. How do you know which link at *D* to erase? When you have erased it, how do you decide which girl to link Bob with next?

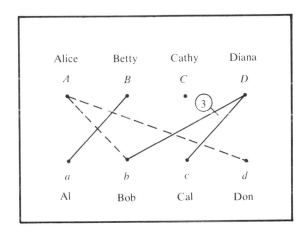

FIGURE 1–2b.

Exercise 1–4. Explain Figure 1–2c, which continues the example.

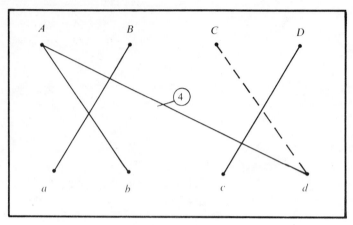

FIGURE I–2c.

In Figure 1–2d you see the final schedule you offer your customers. It is a stable schedule, by the way you constructed it, for you can prove to any doubting patron that there is no unstable situation like that described in Definition 1–1.

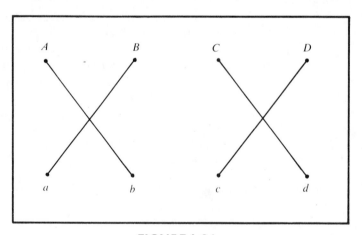

FIGURE I–2d.

You may prefer to replace the blackboard sketches suggested in Figures 1–2a, b, c, and d by a solitaire card game. Label each of four 3 by 5 index cards at the top with one of the girls' names. Then lay these cards in a row. Cut two other index cards in half to make four small cards. Now "play" the boys' cards on the girls' cards, as follows: First, play each boy's card on the card of the girl he ranked first. In the example, then, play Al's card on Betty's, Bob's on Diana's, Cal's on Diana's, and Don's on Betty's. Resolve conflicts (that is, two small cards on the same big card) according to the girl's preference, rejecting the one she ranked lower. Then play the rejected boy's card according to his next choice. Figure 1–2e illustrates the card game at the stage that corresponds to blackboard sketch Figure 1–2c.

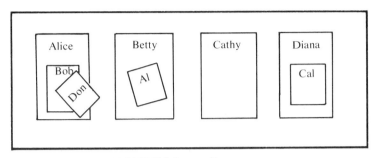

FIGURE 1–2e. CARD GAME.

Theorem 1–1. The process described yields a stable assignment.

PROOF. Suppose one of the boys, Adam, preferred another girl, Eve, to his own date and Eve preferred Adam to her date. Then, by the construction, you tried to match Adam to Eve before you paired him with his present date, but it led to a conflict, and you erased the link from Adam to Eve, because she ranked someone else higher in preference. The date you have assigned her is that preferred one or one who, in a later conflict, rated still higher. ∎*

Notice that you have done no calculations at all in working the example, not even algebraic manipulation of equations. Yet the process is a mathematical one and the explanation of why the solution in Figure 1–2d is stable is a genuine proof. Partly because you are not used to this kind of proof, you may have to read it several times to feel comfortable with it.

It can be proved that Figure 1–2d gives a stable solution that is *optimal* for the boys, in the sense that each boy is at least as well off under the assignment as he would be under any other stable solution. However, this solution is not optimal from the girls' points of view. To optimize the assignments for girls, start by assigning each girl her favorite boy, as shown in Figure 1–3a.

---

* The symbol "∎" means that the proof is complete at this point. In some cases additional commentary may follow, but it is not part of the proof.

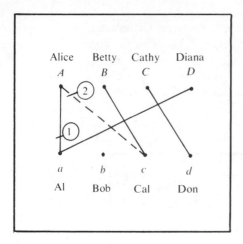

FIGURE I–3a.

> **Definition I–2.** A stable assignment is **optimal** for the applicants whose choices are considered first if each such applicant is at least as well off as under any other stable assignment.

Directly from the definition, you can see that if the first applicants do have an optimal assignment, then they have only one of them.

Exercise 1–5. In Figure 1–3a resolve the conflict over Al by considering his preference between Alice and Diana. What is Alice's next choice? Where does the next conflict occur?

Exercise 1–6. Complete the girls' optimal assignments, as in Figure 1–3b.

Exercise 1–7. Compare Figures 1–2d and 1–3b. Explain why each one represents a stable solution. Which boys are better off in Figure 1–2d than in Figure 1–3b? Which girls are better off in Figure 1–3b?

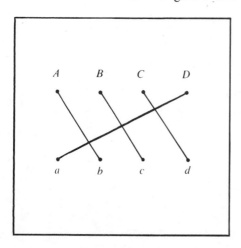

FIGURE I–3b.

Exercise 1–8. Follow the process outlined to get a stable solution optimal for boys for the preference table in Figure 1–4.

		A	B	C	D
			Girls		
Boys	a	1, 2	3, 1	2, 1	4, 4
	b	1, 1	2, 2	3, 2	4, 3
	c	2, 3	1, 4	3, 3	4, 2
	d	2, 4	3, 3	1, 4	4, 1

FIGURE 1–4.   TABLE OF PREFERENCES.

Exercise 1–9. Give the girls first choice in the preferences shown in Figure 1–4 to get a stable solution that is optimal for the girls. Show that in this case the solution for boys in Exercise 1–8 is also the solution for girls.

The process you have used to find a stable schedule in this example is an "algorithm," that is, a step-by-step construction that gives a solution and at the same time proves that it *is* a solution. Long division is an algorithm, and so are procedures for choosing "It" in a game (such as "Eenie, meenie, minie, mo" and cutting for high card); even step-by-step directions for adjusting a carburetor can be considered an algorithm. Although algorithms may lack the elegance of more concise solutions, they have the desirable feature of being constructive. They have increased importance in this age of computers (as illustrated in Chapter 10), for they readily supply the kind of sequenced instructions a computer needs. Consequently, a person who cultivates a taste and talent for inventing algorithms fills a real need. It is pleasing to think that two superb "algorists" of the past, the Swiss mathematician Leonard Euler (1707–1783) and the German mathematician C. G. J. Jacobi (1804–1851), would find their methods in great demand today.

The dating-service example was adapted from a paper by D. Gale and L. S. Shapley, "College Admissions and the Stability of Marriage."*

_____

* American Mathematical Monthly, Vol. 69, No. 1, January, 1962, pp. 9–15.

The "college admissions" or "fraternity rushing" scheme they present is a generalization of the dating, or stability-in-marriage, example:

College Admissions

Suppose each college A, B, C, ... has a quota of freshman students that it can accept. Let each prospective freshman fill out a preference table, ranking the colleges that he will consider choosing, leaving out any that he would not attend under any circumstances. Each student applies to his first-choice college. Any college receiving more applications than it can accept, that is, more than its quota q, puts its favorite q applicants on a waiting list and sends letters of rejection to the others. Students rejected in this first stage apply to their second-choice college. Now each college considers its waiting list plus the new applicants. If the combined set has more than q members, the college selects its favorite q for the new combined list, sending letters of rejection to the others. Rejected students apply to their next choices. The process ends when every student is on a waiting list or else has been rejected by every college that he would consider attending. Each college then accepts those on its waiting list.

Exercise 1–10. Prove that the process outlined leads to a stable solution of the college-admissions problem: if a student s is not accepted by college C but would have preferred college C to the one that accepted him, what can you say about the preference of college C where student s is concerned? (Reread the proof of Theorem 1–1 for suggestions.)

A process like the admissions procedure described by Gale and Shapley is actually in use in the intern matching program, by which interns are placed at hospitals throughout the United States.* Applicants rank hospitals and each hospital ranks all applicants it would consider; then a computer carries out the step-by-step process in a short time.

See whether you can follow this proof that the college-admissions solution is optimal for the students. You may find it harder than it looks. In fact, it is quite optional, included mainly to intrigue you.

> **Theorem 1–2.** The college-admissions process described here yields an assignment that is optimal for the students.

PROOF. Let the assignment yielded by the described process be called S. That is, S lists exactly which students are finally accepted by each college. You want to prove that S is optimal for the students; that is, that every student is at least as well off under assignment S as under any other stable assignment. Suppose this were not so, that there is a student who would prefer to be in a different college. You can prove that any

---

*National Intern and Resident Matching Program, State National Bank Building, Suite 1150, Evanston, Illinois 60201.

assignment T that sends him to that different college is unstable, and hence is not eligible for comparison with the optimal assignment S:

Call a college A "possible" for an applicant a if there is some stable assignment that sends him to A. Then you want to show that if assignment S does not send some student a to A, although a prefers A, then A is impossible for a.

Assumption 1: Suppose that up to a certain stage in the construction of assignment S no student has been rejected by a college that is possible for him. At this stage college A rejects applicant a and puts its quota q of other applicants on its waiting list, $b_1, b_2, b_3, \ldots, b_q$. Each of the students b_i prefers A to every other college that is possible for him, since any colleges that have rejected him were impossible for him at this stage, by Assumption 1. Suppose that some assignment T sends a to college A. Then to keep A within its quota, at least one of the students b_i, say b_1, must go to some other college, say B, under assignment T. But A prefers b_1 to a, and b_1 prefers A to B, yet b_1 does not go to A under assignment T. Then T is unstable. ∎

Exercise 1–11. The mercenary armed forces of a country, the Army, the Navy, and the Marines, compete for recruits among the 15 men listed in Figure 1–5. The first entry opposite a man's name gives his ranking of

	Army	Navy	Marines
Abercrombie	1, *1*	2, *12*	3, *7*
Bradford	3, *9*	1, *11*	2, *6*
Crumcakor	2, *2*	1, *10*	3, *8*
Duffy	2, *10*	3, *13*	1, *1*
Edgemont	3, *3*	1, *9*	2, *5*
Fitch	2, *11*	1, *8*	3, *9*
Green	2, *4*	1, *7*	3, *10*
Hale	1, *5*	3, *14*	2, *4*
Irving	3, *6*	2, *6*	1, *2*
James	1, *12*	2, *5*	3, *11*
Knowles	2, *15*	1, *4*	3, *12*
Lewis	2, *13*	1, *3*	3, *13*
Montcalm	1, *7*	2, *2*	3, *14*
Norris	3, *8*	2, *15*	1, *3*
Oppenheim	2, *14*	1, *1*	3, *15*

FIGURE 1–5. ARMED FORCES PREFERENCE TABLE.

the force listed in the vertical column; the second entry shows how the force ranks the man. The Army has a quota of 8, the Navy 4, and the Marines 3. Have each of the armed forces make an offer to its quota of men, starting with those it likes the best. Let each man reject all but his favorite one among those making him offers, and ask the favorite for time to consider. Then let each military service that has been rejected apply to new men. Show the final assignment derived by the college-admissions procedure.

Exercise 1-12. Compare the situation in Exercise 1-11 to that of industries trying to attract new employees.

A TRANSPORTATION PROBLEM

A national student organization has 10 resource leaders in its East Coast office and 7 in its West Coast office. The organization is to hold three simultaneous regional planning conferences. The conference at Reno, Nevada, is to have 4 of the resource leaders in attendance, the conference at El Paso, Texas, is to have 8, and the conference at Elkton, Maryland, is to have the other 5. Figure 1-6 shows the cost, in hundreds of dollars, of sending a resource leader to the respective conferences (cost includes transportation, per diem, and so forth).

	Reno	El Paso	Elkton
East office	8	6	1
West office	1	2	8

FIGURE 1-6. COST TABLE.

The problem is how to allocate the 17 leaders so as to minimize costs.

You can solve the problem in steps, first writing a **feasible** solution, that is, one that satisfies the conditions on the number of leaders available from the East and West offices and the numbers needed at the respective conferences. Temporarily delay the problem of minimizing cost. Arbitrarily taking the East office's 10 leaders first, just because they happen to be listed first, supply the needs at Reno, that is, supply 4 from East to Reno. As there are 6 leaders left at the East office, let them go to El Paso.

This leaves El Paso with a need for 2 more leaders, so supply them from the West office. The remaining 5 at the West office go to Elkton.

In the allocation problems we shall study here, the total supply always equals the total demand exactly, and a feasible solution merely satisfies the requirements on totals at each supplier (East 10 and West 7) and at each demander (Reno 4, El Paso 8, and Elkton 5).

Figure 1–7 sums up this first feasible solution, and also records the cost in hundreds of dollars per person for each link.

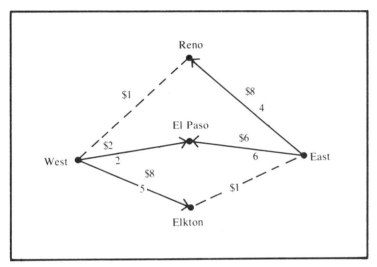

FIGURE 1–7. A FIRST FEASIBLE SOLUTION.

There is only a dotted link from East to Elkton, since the first feasible solution had East supply none of the Elkton positions. Suppose now we try sending someone from East to Elkton. The new link completes a loop, East-Elkton-West-El Paso-East. Any increase along the link East-Elkton must be equalized by a decrease somewhere in the loop. In fact, if East sends one leader to Elkton, that makes one fewer that it can send to El Paso. The extra person at Elkton results in a decrease of a leader from West to Elkton, but then West has an additional leader to send to El Paso. Figure 1–8 shows the loop with the increases and decreases.

What change in costs will there be for each person sent from East to Elkton? The added person from East to Elkton costs $1 (in hundreds). The person no longer needed from West to Elkton decreases the cost by $8. West to El Paso adds $2, and East to El Paso subtracts $6. The net change along the loop is

$$+\$1 - \$8 + \$2 - \$6 = -\$11, \text{ a decrease.}$$

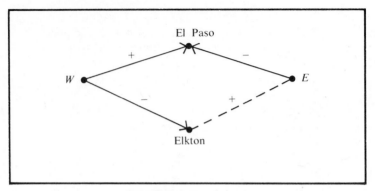

FIGURE I–8. A LOOP.

Since the organization wants to minimize costs, the change represents an improvement, an improvement that can be made five times before the link West-Elkton disappears; that is, before all 5 people from the West-Elkton link have been diverted to El Paso. Figure 1–9 shows this change.

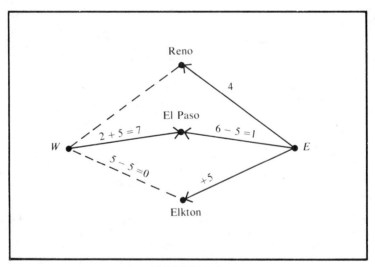

FIGURE I–9.

Now suppose West supplies someone to Reno. Figure 1–10 shows the resulting loop and change in costs. Since this change decreases costs, make the change four times until the East-Reno link disappears because all 4 people from the East-Reno link of Figure 1–9 have been diverted to El Paso. Figure 1–11 shows this change.

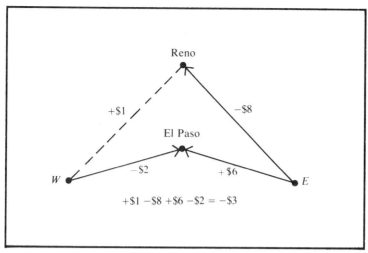

FIGURE I-10.

Increasing either of the dotted links from zero in Figure 1–11 would complete a loop already optimized, so Figure 1–11 represents the least expensive allocation of leaders. Notice that it conforms to the common-sense rule of thumb that it will be cheapest to supply western conferences mostly from the West coast office and eastern conferences mostly from the East coast office.

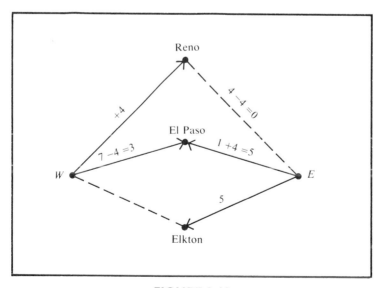

FIGURE I-11.

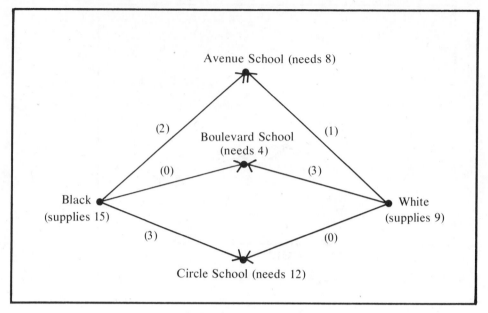

FIGURE 1-12. SCHOOL EXAMPLE.

EXAMPLE   Annexation has brought 15 new black pupils and 9 new white pupils into a school district of 3 schools. Avenue School can take 8 more students, Boulevard School can take 4, and Circle School can take 12. Figure 1–12 gives indices representing the social utility attached to sending pupils to the various schools, based on the racial percentage already

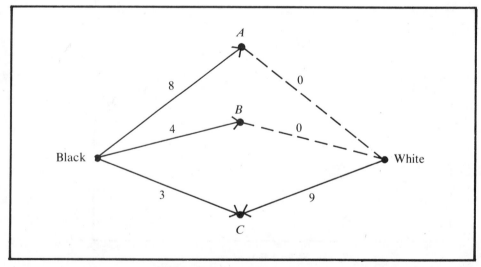

FIGURE 1-13. INITIAL FEASIBLE SOLUTION.

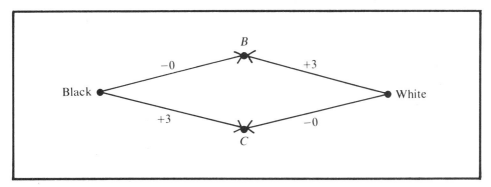

FIGURE 1–14.

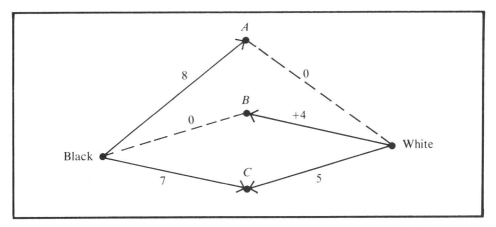

FIGURE 1–15.

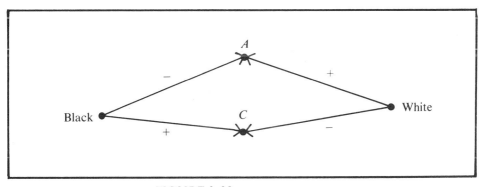

FIGURE 1–16.

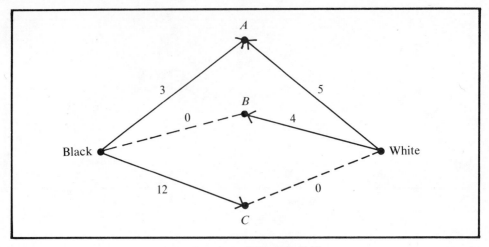

FIGURE 1-17.

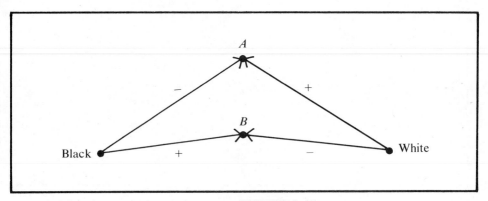

FIGURE 1-18.

existing in each school. The index is high for a given race-school combination if the percentage already at the school is low. Figure 1–13 shows a feasible solution to start with. Since the loop B-White-C-Black-B results in a net social improvement of $+(3) - (0) + (3) - (0) = +(6)$ index points, make the change until the loop breaks at Black-B, that is, make the change 4 times. Notice that the student-organization problem called for minimizing cost, making decreases desirable. Here you are trying to maximize a social index, making increases desirable.

If you increase the link White-A from zero, you complete a loop White-A-Black-C-White. The change in social index is $+(1) - (2) + (3) - (0) = +(2)$, an improvement, so make the change until link White-C decreases to zero, that is, 5 times.

An increase in the link Black-B completes loop Black-B-White-A-Black. The resulting social change is $+(0) - (3) + (1) - (2) = -(5)$.

Since this is negative it results in a worse social picture, so leave out the link Black-B.

The only other possibility is to increase the link White-C, but you determined that it was better socially to send those students to A. Then the optimal solution is the one shown in Figures 1–17 and 1–19.

Notice that these transportation problems call for making the total supply equal the total demand, while minimizing or maximizing some index such as cost.

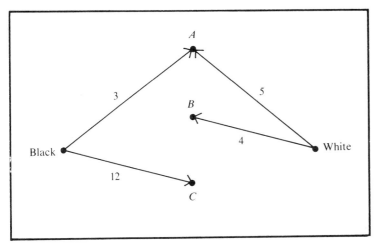

FIGURE 1–19.

Exercise 1–13. You have 5 evening hours E and 3 early morning hours M to devote to studying math-and-science courses S and nonscience courses N. You need to spend 4 hours on science courses S and 4 hours on nonscience N. You have found that you perform better on S homework when you are fresh in the morning, and that you prefer to write essays and do reading assignments for N in the evening. You have expressed the utility of hours for courses as in Figure 1–20. Find the best way to allocate your study time.

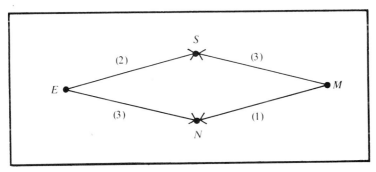

FIGURE 1–20. ALLOCATION OF HOMEWORK TIME.

Exercise 1–14. You are an efficiency expert, hired to streamline factory operations. One assembly operation calls for three kinds of large parts, P, Q, and R, which are to be stocked on two shelves, s and t. Shelf s will hold 10 parts; shelf t will hold 11 parts. There are 7 parts P, 7 parts Q, and 7 parts R to be stocked at a time. Figure 1–21 shows an index of the energy necessary to lift·a part from the storage shelf to the place where it will be used. Find the best way to distribute the stored parts between the two shelves.

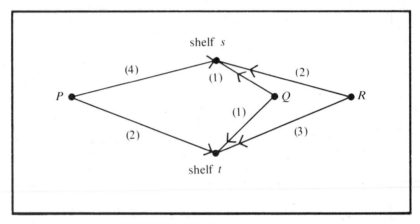

FIGURE 1–21. STORAGE OF PARTS.

Exercise 1–15. In a freshman class, 49 mathematically disadvantaged students D and 260 well-prepared students W are to be assigned to three math lecture sections of about 103 students each. Figure 1–22 gives an index of how well the three lecturers, Faith, Hope, and Charity, seem to succeed with students of various preparations. How should the assignments be made? To bring this exercise nearer reality, you are

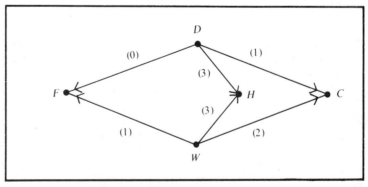

FIGURE 1–22. CLASS ASSIGNMENTS.

encouraged to unbalance the three class sizes slightly, at your discretion, but not by more than a discrepancy of, say, 5 students, which might cause dissension on the faculty.

Exercise 1-16. There are 16 people to be assigned to three advisory boards, a Human Relations Board, a Recreation Board, and a School Study Board. The H.R. Board needs 5 people, the Rec. Board needs 3, and the S. S. Board needs 8. It has been determined that of the 16 people, 6 are welfare recipients and 10 are not. Figure 1-23 gives indices representing the desirability of placing people on the respective boards. Optimize the assignment by the network method explained in this section.

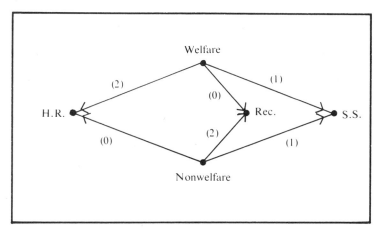

FIGURE 1-23. ADVISORY BOARDS.

Then show how the solution agrees with the common-sense solution you would have tried before you had heard of this method. What special feature of the weights (indices) in Figure 1-23 makes solution easy in this case? Return to some of the previous exercises and discuss the given indices, mentioning examples in each case to show that the index size can be "out of order" so as to make it initially unclear how to assign the sets optimally.

Exercise 1-17. You are a fowl fancier and raise 10 chickens, 6 ducks, and 8 geese. You have room for 11 fowl in batteries and 13 fowl in pens. How would you allocate the fowl to maximize your return if the diagram indicates the profitability based on better price for clean eggs from battery fowl but higher cost for battery fowl. (See Figure 1-24.)

The diagrams you have been drawing for transportation problems can help you become objective about analyzing your own thought processes. We hope you do not always think in circles; but notice that when

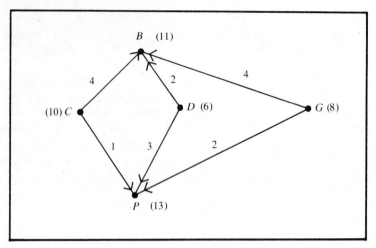

FIGURE I–24. HOUSING FOWL.

you are faced with an allocation problem, you do think in loops. "I have only so much money for clothes, so if I buy that suit I'll have to pass up the shoes. If I pass up the shoes, I'll save enough to . . . ." Loop thinking is so automatic as you allocate limited funds, limited time, and limited space that you may never have externalized it enough to realize you were doing it. The diagram form cuts down on errors like neglecting a possibility and, like most systematizations, adapts your thought form for machine use, as for a computer.

A good reference for more on the transportation problem and other interesting problems is *Finite Mathematics*, by Guillermo Owen (W. B. Saunders Company, 1970).

GRAPHS

In the transportation problem you used some techniques from the theory of "graphs." Graphs are simple diagrams for recording relationships. The range of the word "relationship" is tremendous, and takes us immediately to the center of sociology, anthropology, logic, traffic control, political science, or electrical circuitry, depending on which relationships are meant.

Definition I–3. A **graph** is a set S, conveniently represented by a set of points

$$S = \{x, y, \ldots, u\}$$

together with a relation V giving for some members x of S a related member y. In a graph diagram a directed **arc** from x to y shows that x has the related member y, and the relation V as a whole is represented by the set of (directed) arcs.

In Figure 1–25 the set S is made up of the 8 points x, y, z, r, s, t, u, w. **EXAMPLE**
The relation V is made up of the 7 arcs $(xy), (yw), (wy), (uz), (sz), (sr), (ru)$.

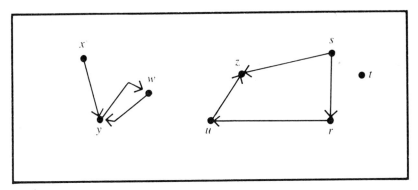

FIGURE 1–25. AN 8-POINT GRAPH.

In Figure 1–26 the set S is $\{x, y, z, u, w\}$. The relation V is **EXAMPLE**
$V = \{(xy), (xz), (yw), (wu), (uz)\}$.

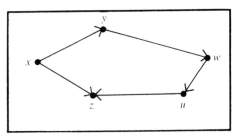

FIGURE 1–26. A 5-POINT GRAPH.

Definition 1–4. In a graph a **path** from x_1 to x_j is a sequence
of arcs in V

$$(x_1 x_2), (x_2 x_3), \ldots, (x_{j-1} x_j),$$

each arc after the first beginning at the endpoint of the preceding.
The path can be written

$$(x_1 x_2 x_3, \ldots, x_j).$$

For some purposes the arcs are considered without noting their
direction, in which case they are properly called **edges**. An edge
can be shown with arrow points at both ends or neither end. A
path of (undirected) edges is called a **chain**.

EXAMPLE   In Figure 1–25 there is a path (*xyw*) from *x* to *w*, but no single arc from *x* to *w*. There is a path from *s* to *z*, (*sruz*), but no path from *z* to *s*.·
However, there are two chains from *z* to *s*, (*zs*) and (*zurs*).

Exercise 1–18. Show that the graph in Figure 1–26 has a path from *x* to every other point of the graph. Show that there is no path, however, from any other point to *x*. Are there chains to *x* from every other point?

> **Definition I–5.** A graph is **strongly connected** if for every two points *x*, *y* of *S* there is a path from *x* to *y*. A graph is **weakly connected** if there is a chain between every two points.

EXAMPLE   From Exercise 1–18 the graph in Figure 1–26 is not strongly connected, since there is no path from *y* to *x*, for instance. It is weakly connected, since there is a chain from every point to every other point.

Exercise 1–19. Consider Figure 1–7 as a graph with the solid lines representing the edges (undirected) of *V*. Find a chain from Reno to West. Show that the graph is weakly connected.

FIGURE I–27. ONE-WAY STREET GRID.

Consider a graph as representing streets, some of them restricted by one-way signs. If motorists can drive from each point to each other point without disobeying the signs, then the graph is strongly connected. If pedestrians, who are not affected by the one-way signs, can walk from each point to each other point, then the graph is weakly connected.

Exercise 1–20. Experiment with 3-point graphs until you can completely characterize a strongly connected 3-point graph. What requirement can you impose to be sure the graph will be at least weakly connected? Remember that both (*xy*) and (*yx*) can be arcs in *V*.

Exercise 1–21. Show that the graph of Figure 1–28 is weakly connected. Show that by adding one more arc you can make it strongly connected.

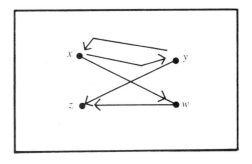

FIGURE I–28. Connectedness.

If the graph of Figure 1–28 represents a traffic problem, you can see that traffic is 2-way between *x* and *y*, one-way from *x* to *w*, one-way from *w* to *z*, and one-way from *y* to *z*. If the city commission decides to make traffic 2-way between *y* and *z*, then you can drive from *z* to *w* by the route or path (*zyxw*).

Suppose that the graph of Figure 1–28 has been used to record a relationship in a family—say, the relationship "gives advice to." If *x* is mother, *y* father, *z* daughter, *w* son, then the mother and father give advice to each other, the mother advises the son, the father advises the daughter, and the son advises the daughter. If the daughter succeeds in giving advice to the father (extending the relationship recorded in Figure 1–28), then he can pass it to the mother, who can in turn advise the son.

Suppose the graph of Figure 1–28 represents pecking orders in a barnyard. Hens *x* and *y* peck each other, *y* pecks *z*, *x* pecks *w*, and *w* pecks *z*. Every hen is pecked by some other hen, but there is one hen that pecks no other hen.

Exercise 1–22. Interpret the graph of Figure 1–28 for plane geometric figures, *x* = a square, *y* = an equilateral rectangle, *z* = a parallelogram, *w* = a rectangle, and let a directed arc represent "is necessarily"; for instance, let *w* → *z* represent "A rectangle *is necessarily* a parallelogram." For this example what other arc would be appropriate?

Exercise 1–23. You are observing a nursery school play yard as part of a study of child development. During the first week, your records show that Josie has played with everyone except Eileen, Gretschen has played with all the boys, David has played with all the boys, May and Jeannette have played only with each other and with Josie, and Steve has played with Harvey. Draw a graph to summarize this information. Undirected edges would be appropriate, unless you are interested mainly in which children initiate each play episode. Use your graph to point out other edges that are implied but not specifically mentioned in your notes.

4-15. Symbol-and-Arrow Diagrams of Complex Systems

We are now prepared to consider how the general concepts developed above can be applied to real biological systems, where the variables of interest rarely, if ever, occur in isolated pairs but rather as part of a great complex of variables bound together by an intricate network of functional relationships. Figure 4-4 illustrates such a complex. The system depicted

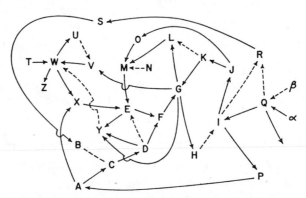

FIG. 4-4. The Complex Relationships among the Variables of a Real Biological System
Each letter in this diagram symbolizes a variable important in the cardiovascular system. It is rarely possible to subject so complicated a system *in toto* to mathematical analysis. The biological scientist must usually content himself with tackling subsystems, isolated in the manner described in Section 4-16.

in Figure 4-4 was not, as the reader might be tempted to suppose, cunningly devised by the author as an impressively bewildering but wholly imaginary example. In fact, Figure 4-4 is a diagram (already considerably simplified) which originally showed some of the actual interrelationships among various factors which influence cardiovascular function. But since we are at present interested only in how to deal with complex systems in general, the real variables of the system (heart rate, coronary blood flow, peripheral vascular resistance, and the like) have been arbitrarily represented by letters of the alphabet.

FIGURE 1–29.

This application of a directed graph is in the field of physiology, but notice how much it has in common with the example of child development in Exercise 1–23. (From *Control Theory and Physiological Feedback Mechanisms*, by Douglas Shepard Riggs. © 1970, The Williams & Wilkins Co., Baltimore.)

Definition 1–6. An **algebraic graph** is a graph showing two different relationships for the same set S of points. One relationship is called "positive" and drawn as solid arcs; the other is called "negative" and drawn as dotted arcs.

Algebraic graphs are useful for showing a positive relationship P, such as "likes," and a negative N, such as "dislikes."

Figure 1–30 shows in solid arcs a relation P "likes," and in dotted **EXAMPLE** arcs a relation N "dislikes" in a business meeting. Is there anyone nobody else likes? Does everyone at the meeting actively dislike him?

Does anyone actively like everybody else? Does anyone hold no active dislikes? Does anyone like someone who dislikes him?

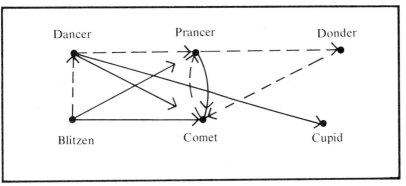

FIGURE 1–30. AN ALGEBRAIC GRAPH.

Exercise 1–24. Represent the following information by an algebraic graph with undirected edges: On a double date, Sue and Ted are attracted to each other, Ted and Alice are attracted to each other, Sue and Mike are attracted to each other, and Alice and Mike are attracted to each other. However, the men find each other repugnant, and the girls find each other insufferable.

The graph of Exercise 1–24 has an edge between every two points, which in the illustration means that each person is linked either positively or negatively with every other person; no relationships are neutral. Such a graph is called "complete." In the discussion of balance in graphs we are going to narrow our attention to complete, symmetric algebraic graphs.

Fortunately, there is a one-word substitute for this list of modifiers:

> **Definition 1–7.** A **signed graph** is a symmetric (undirected) algebraic (two relations) graph that is **complete** (every point related to every other point).

The question of whether a signed graph appears balanced is suggested by generalizations about equilibrium in some human relationships, such as:

"My friend, any friend of yours is a friend of mine."
and
"They have a lot in common; they both hate the same people."

For signed graphs of just three points, there are only four different structural possibilities, as shown in Figure 1–31. All three relationships may be positive, two may be positive and one negative, one may be positive and two negative, or all may be negative. If we take the positive relationship to be friendship and the negative to be enmity, triangle (a) shows three friends. In triangle (b), known as the "eternal" triangle, x and y are friends and x and z are friends, but y and z are enemies. In triangle (c) y and z are friends, but neither gets along with x. In triangle (d) the enmity is general.

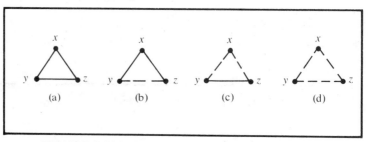

FIGURE 1–31. SIGNED GRAPHS HAVING THREE POINTS.

Triangle (b) with one negative relationship and two positive relationships tends not to remain stable. It is said to be not "balanced," since it tends toward some other arrangement. Of course, this is a whopping social generalization. However, it turns out to produce helpful tools if we define "balanced" in conformity to this idea. Triangles (a) and (c) are considered to be balanced—everyone happy in (a), and in (c) x disliking y and also y's friend z, y and z as friends both disliking x, y disliking x and so drawn to z who also dislikes x, and so on. Triangle (d) is considered not balanced, as two enemies may at any time decide that their hatred of the third is more intense than their hatred of each other and so form a coalition against the third.

> **Definition 1–8.** The sign of an edge in the relationship P is taken to be $+$; the sign of an edge in N is $-$. To find the **sign of a chain** in a symmetric algebraic graph, form a product of $+1$'s and -1's, the signs agreeing with those of the edges that make up the chain. The chain is $+$ if the product is $+1$, otherwise $-$.

> **Definition 1–9.** In a graph a path from x to x is called a **circuit**. If the graph is symmetric (undirected), a chain from x to x is a **cycle**.

> **Definition 1–10.** A cycle of length 3 is **balanced** if its sign is $+$; it is not balanced if its sign is $-$.

Exercise 1–25. Check each of the triangles in Figure 1–31 for balance. Observe that the results follow the personal situations described.

> **Definition 1–11.** A signed graph is a **balanced graph** if all of its triangles are balanced.

> **Theorem 1–3.** A signed graph is balanced if and only if its points S can be separated into two distinct parts T and U, where every edge between two points of T is positive, every edge between two points of U is positive, and every edge between a point of T and a point of U is negative.

PROOF. First, suppose $S = T \cup U$, with $T \cap U = \varnothing$. This set language (see Chapter 4)* says that S splits into two distinct parts. Suppose

---

* Cross references are included to show how the chapters are interrelated. You may want to skip some references and exercises until you have read the related chapters.

that each internal edge of T is positive and that each internal edge of U is positive, but that each edge between T and U is negative, as in Figure 1–32.

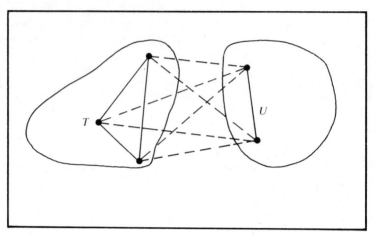

FIGURE I–32. ILLUSTRATION FOR THEOREM 1–3.

You want to prove that S is balanced, which by Definition 1–11 means that all its triangles have sign $+$. If all the points of a triangle lie in T, then by the hypothesis all the edges are $+$, so the sign of the triangle is $+$:

$$(+1)(+1)(+1) = +1.$$

Similarly, if all points of a triangle lie in U, then all the edges are $+$, and so is the triangle. Now take a triangle that has an edge (tu) with t in T and u in U. Its third point v must be either in T or in U, since S splits, by hypothesis. If v is in T, then edge (tv) is $+$ and edge (vu) is $-$, so that for the triangle

$$(-1)(+1)(-1) = +1$$

the sign is $+$.

Exercise 1–26. Show what happens if v lies in U.

Now prove the theorem in the other direction; that is, suppose that the graph is balanced and prove that its points S separate into distinct subsets as stated in the theorem. Let x be some point of S. Form the set T of all points y of S for which (xy) is a positive edge. (Use the fact that the graph is complete, so that every two points have some connection, either $+$ or $-$.) Let all other points of the graph constitute the set U. If the set U is empty, that means that all edges of the graph are $+$, and the theorem is satisfied trivially. Let u be a point in U. Then (xu) is negative, since u is not in T. Similarly, if w is another point in U, (xw) is negative.

Then from the hypothesis that every triangle is balanced, triangle (suw) is positive, so that (uw) must be positive for every pair in U.

If x and y are in T and u is any point of U, then (xu) is negative and (xy) is positive, so any edge (yu) with y in T and u in U is negative, to balance the triangle (xyu). ∎

Theorem 1–4. A signed graph is balanced if and only if all of its cycles are positive.

PROOF. If all the cycles are positive, then certainly all the cycles of length 3, or triangles, are positive, so the graph is balanced.

To prove the "only if" part of the theorem, assume that the graph is balanced. Then by Theorem 1–3 separate its points into distinct sets T and U, with edges interior to T or interior to U positive, and edges between T and U negative. Any cycle of the whole graph remains entirely within T or entirely within U, in which case all its edges are positive and the cycle positive, or else the cycle crosses from T to U. However, a cycle returns to its starting point, and that starting point is arbitrary. If the cycle crosses from T to U, then it must return to T eventually to close the cycle. In fact, each edge (tu) of the cycle must be matched by some edge $(u't')$ with u' in U, t' in T, to close the cycle. Thus, every cycle is made up of interior edges, which are $+$, and pairs of exterior edges, which contribute $(-1)(-1) = +1$ to the sign of the cycle. ∎

Theorem 1–3 brings to light an important fact that you can infer from the definition of balance; namely, that a division into more than two blocks is not balanced, or stable, and will tend toward a two-block situation. For instance, at a party where politics is discussed, if each person is a partisan Democrat or a partisan Republican, sentiment is not likely to remain divided into 3 or more blocs. A situation like Southern Democrat, other Democrat, Republican, may lead to a Southern Democrat-Republican coalition, a uniting of both Democrat blocs, or a coalition of Republicans and non-Southern Democrats against the Southern Democrats.

Exercise 1–27. Draw a signed graph having 6 points. To make it complete you will need 15 edges. Let 8 edges be negative and make the graph balanced. Find the two sets T and U of Theorem 1–3.

In Figure 1–33 you see reproduced a page of the "Girl Scout Leader Notebook," giving a process for dividing a troop of Girl Scouts into patrols. The process described does not fit exactly any of the graphs you have studied, but it exhibits many of the same features. Unlike the college-admissions process, the Girl Scouts' method does not always yield a solution. The following exercises will help you discuss the process in the light of elementary graph theory.

FORMING PATROLS BY WRITTEN CHOICE

1. ASK EACH GIRL TO WRITE NAMES OF THREE OTHERS:

Estelle	Jackie, Lillian, Nancy
Florence	Sophie, Dot M., Eleanor
Peggy	Eleanor, Eileen, Dot M.
Jackie	Lillian, Dot M., Grace
Dot R.	Dot M., Sally, Jackie
Nancy	Dot R., Robin, Estelle
Robin	Nancy, Sally, Kate
Mary	Sue, Betty, Jane
Sue	Mary, Betty, Jane
Betty	Mary, Sue, Jane
Jane	Sue, Betty, Mary
Pat	Lillian, Sally, Estelle
Norma	Eleanor, Lillian, Eileen
Lillian	Eleanor, Dot M., Leona
Sally	Eleanor, Nancy, Robin
Eleanor	Sally, Barbara, Lillian
Dot M.	Sally, Dot R., Florence
Eileen	Peggy, Kate, Eleanor
Kate	Eileen, Peggy, Florence
Helen	Grace, Lillian, Dot M.
Grace	Lillian, Helen, Jackie
Leona	Sally, Lillian, Betty
Sophie	Florence, Dot M., Eleanor
Barbara	Eleanor, Peggy, Helen

2. LIST THE CHOICES EACH RECEIVES

Estelle	Florence	Peggy	Jackie
Pat	Kate	Kate	Grace
Nancy	Sophie	Barbara	Estelle
	Dot M.	Eileen	Dot R.

Dot R.	Nancy	Robin	Mary
Dot M.	Sally	Sally	Sue
Nancy	Estelle	Nancy	Betty
	Robin		Jane

Sue	Betty	Jane	Pat
Mary	Mary	Mary	
Betty	Sue	Sue	
Jane	Jane	Betty	
	Leona		

Norma	Lillian	Sally	Eleanor
	Pat	Pat	Norma
	Norma	Eleanor	Lillian
	Eleanor	Dot M.	Sally
	Helen	Leona	Eileen
	Grace	Dot R.	Sophie
	Leona	Robin	Barbara
	Estelle		Florence
	Jackie		Peggy

Dot M.	Eileen	Kate	Helen
Lillian	Norma	Eileen	Grace
Helen	Kate	Robin	Barbara
Sophie	Peggy		
Florence			
Peggy			
Jackie			
Dot R.			

Grace	Leona	Sophie	Barbara
Helen	Lillian	Florence	Eleanor
Jackie			

3. CONSIDER FACTORS SUCH AS THESE:

Eleanor, Dot M., Lillian, and Sally, with 6 to 8 choices each, are obviously popular and may be natural leaders. Place one of these in each patrol.

No one chose Pat or Norma. Be sure they get their first choice and at least one other. Leona, Sophie, and Barbara were chosen only once each. Try to give these girls two of their choices.

Mary, Sue, Betty, and Jane chose only one another. They need to widen their friendships. Split them in twos.

Several girls chose each other. Place them in the same patrol if possible.

4. FORM THE PATROLS

Patrol 1	Patrol 2	Patrol 3	Patrol 4
Lillian	Sally	Eleanor	Dot M.
Pat		Norma	
Estelle		Eileen	
		Barbara	
		Peggy	
	Leona		Sophie
			Florence
	Betty		Mary
	Jane		Sue
Helen	Grace	Kate	Dot R.
Robin	Jackie		
Nancy			

Exercise 1–28. Is the situation symmetric, or would you need directed arcs to describe it?

Exercise 1–29. Is it strongly connected? Is it weakly connected?

Exercise 1–30. How could you express in set language (Chapter 4)* the fact that the troop is to be separated into distinct patrols, with every girl in some patrol?

Exercise 1–31. In graph theory a "clique" is defined to be a subgraph whose points S' form a subset of S and whose arcs form a subset V' of V, such that arc (xy) is in V' if and only if x and y are both in S'. Find a clique in the Scout example.

Exercise 1–32. Borrow vocabulary from the college-admissions section to describe as accurately as possible what the objectives of the Scout process are.

Exercise 1–33. Show that cases could arise that would not lead to a solution by the Scout process, in the sense that some of the objectives could not be met. For this kind of analysis, turn to "trivial" cases. Suppose, for instance, that one girl is new to the community and to Scouts. The most popular girls are the only ones she has heard of by name, so all of her three choices are among the four most popular girls. Since none of the girls knows her, none lists her. Can she be included with two of her choices? What adjustments could you recommend for the rules of the process to meet this problem? Leaving mathematical graphs aside for the moment, turn to the actual social situation: What purely social techniques could you use to prevent such cases from arising to complicate your mathematical analysis?

Exercise 1–34. The Scout process refers to "first" choices, yet does not have the girls rank their three choices. Discuss social grounds for not recommending ranking. Discuss how to allow for this in the process.

Exercise 1–35. Discuss adjustments that might be made if the troop is small.

Exercise 1–36. Shift Helen, Robin, and Nancy to Patrol 2, and Leona, Betty, and Jane to Patrol 1. Which links are gained and which are lost? Discuss the effect of giving Lillian and Sally their choices among the less popular girls, Leona and Robin.

---

* Use notation like that in the first sentence of the proof of Theorem 1–3, or skip this exercise until you have read Chapter 4.

Exercise 1–37. Suppose a clique of four girls is placed in a patrol of six girls. Discuss the balance the patrol may tend toward socially, based on Theorem 1–3.

Exercise 1–38. Show how you can borrow from the transportation problem when applying the Scout process: For instance, you might assign values indicating how important you believe it is to assign girls *g* and *h* to the same patrol. Then as you sort index cards to represent trial assignments, you could calculate the net value of moving card *g* out of one pile into another. How would you calculate this value? When would it be positive? When negative?

With several analyses behind you, you may want to have a class discussion on social situations, relating them to graph theory. If you find graph theory attractive and think it may offer useful features for your main fields of interest, read *Applications of Graph Theory to Group Structure* by Claude Falment (in the Prentice-Hall Series in Mathematical Analysis of Social Behavior, 1963), and perhaps some of the bibliography cited there. You will find "A Variety of Interesting Applications" in the chapter of that name in *Finite Graphs and Networks*, by R. G. Busacker and T. L. Saaty (McGraw-Hill, 1965). The reference lists at the ends of chapters range over many fields of interest. The first chapter of *Graph Theory* by Frank Harary (Addison-Wesley Publishing Company, 1971) shows how graph theory came to be rediscovered many times in different contexts. It is this uneven historical development that accounts for discrepancies in vocabulary from one reference to another.

REVIEW OF CHAPTER 1

1. The dating-service example and the college-admissions example involve **applicants** applying for **partners.** (The "partner" for an applicant for college admission is a college.) When is an assignment of each applicant to a partner a stable assignment? Describe the steps in reaching a stable assignment.

2. What is an **optimal** assignment? Does it make any difference whether the applicants choose first or the partners choose first?

3. A transportation problem involves allocating the total supply (of money, time, and so forth) among several demands. How much must the total demand be? What is a **feasible** solution?

4. List some quantities that might be minimized or maximized in a transportation problem. How do you determine from links whether a feasible solution optimizes the quantity? What adjustments do you make if it does not?

5. Sketch graphs to illustrate each of the following:

arc	edge
path	chain
circuit	cycle
strongly connected graph	weakly connected graph

6. What is an algebraic graph? What is a signed graph?

7. How do you find the sign of a chain in a signed graph?

8. Draw examples of cycles that are balanced and cycles that are not balanced.

9. When is a signed graph called "balanced"?

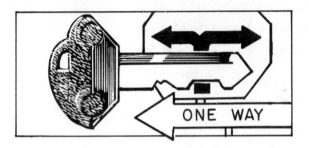

CHAPTER I AS A KEY TO MATHEMATICS

This chapter has been included especially to broaden your view of what constitutes mathematics. It is important to a modern conception of mathematics to see it as more than computation, more than formulas, more than the servant of science.

Beyond these reasons, we had an even more compelling one for including graphs and networks to give you a key to mathematics. We want to put you "in the driver's seat" or, perhaps, to help you realize that you have been there all along! Too often we think of mathematical techniques as ancient and fixed, hard to understand and use, and invariable. We have tried to help you adapt basic ideas to brand new material in new ways, ways that reveal the compromises and value judgments (as in the Scout example) natural to real applications.

The late Professor C. H. Yeaton used to say that the scientific method is "try it and find out." The dating service example and the transportation problem both illustrate the value of this method. Both call for trial solutions at each stage, with a plan to follow in case the trial solution is not good enough. In our grandparents' school days, students were taught to attack almost all arithmetic problems this way: To find out how much one pencil costs if 20 of them cost $1, try 10¢. 20 × 10¢ is too much, so try 2¢, and so forth. Now that computers can carry the computational burden, the try-it-and-find-out approach is really coming into its own. Most of us find it easier to understand than a more formal approach, perhaps because we have had so much practical experience with it. (If I buy this item, I will have just $5.19 left, which would mean I could not buy that item.)

Another feature of these techniques that helps underline your own entrepreneurship is their lack of completeness. There are many problems left to be solved, many extensions of existing techniques to be developed. For instance, there is the "traveling salesman" problem. A salesman wants to cover n cities (the points of the graph), each pair of which is connected, in the most efficient manner, not retracing any part of the route and not revisiting any city, and he wants this simple circuit to bring him back to his home city at the last step. Only special cases of this simply explained problem have been solved. It would be of considerable practical interest if we could develop an algorithm to construct the best route for any graph or to show that the graph has no simple circuit that touches every point.

One student, baffled by the idea of original, invented mathematics, asked, "How can a Ph.D. candidate write an original dissertation in mathematics? Does he have to sit around until something like the Pythagorean theorem occurs to him, or a Newtonian apple falls on his head?" We hope this chapter has suggested to you how a student, possibly taking broad generalizations from nature as a point of departure, can pose interesting problems and innovative solutions.

CHAPTER 2

Rich man, poor man, beggar man, thief,
Doctor, lawyer, merchant, chief,
Tinker, tailor, cowboy, sailor.
Children's Fortunetelling Rhyme

In this chapter we explore a special simplified arithmetic, and in
so doing learn something about arithmetics in general.

CONGRUENCE ARITHMETIC
Arithmetic of Rhythms

You are an agricultural consultant, helping with pig-management projects. The sow in one small village will be in oestrus (heat) on Saturday, and you could arrange to have her bred then, but you want to be sure that the farrowing (births) will be on a weekend when you can arrange to be present. The oestrus cycle of the pig is 21 days, and the gestation period is 112 days. If the sow is bred on Saturday, will she farrow on a weekend? If the sow is not bred until her next oestrus period, will she farrow on a different day of the week?

Cyclical, or periodic, events surround us from our very conception, from the periodic motions of our universe of stars and planets to the seasonal or daily variations in plant and animal life. The arithmetic designed especially for analyzing periodic events is called "modular" arithmetic.

Since a periodic event repeats the same cycle again and again, all you need to analyze is a typical cycle. This is reflected in the fact that modular arithmetic can be encompassed in a finite range, from 0 through 111, say, or from −3 to 4, with no need for any really big numbers at all! Since the mechanics are easy, you have a chance to get above them and understand the workings of the system as a whole. ·This in turn highlights for you the features of ordinary arithmetic.

Exercise 2-1. Suppose that in a certain year the 11th of March falls on a Friday. Find which other March dates of that year fall on Friday.

Exercise 2-2. It is now 3 o'clock. What time was it 12 hours ago? What time will it be 48 hours from now?

Exercise 2-3. It is now 3 o'clock. Write two different time intervals from now when it will be 5 o'clock. Write two different time intervals past (such as 10 hours ago) when it was 5 o'clock.

FIGURE 2–1. Spiral staircase.

In climbing this spiral staircase you would return **periodically** to the same place relative to the ground, but on a higher level.

Exercise 2–4. A simple electrical switch has two positions, closed and open. At present it is open. Suppose you switch it three times. Then its position will be closed. What will its position be after five switches? Describe all the numbers n such that n switchings yield the position "closed." What numbers n yield the position "open"? Can you interpret a negative n?

Exercise 2–5. In Figure 2–2 a wheel rolls at a constant rate and the height of a black point on the rim is recorded for each time. Since the hub line is taken to be "zero" height, the height is recorded as negative whenever the black point is lower than the hub. Find all the times shown for which the height is $\frac{1}{2}$, and forecast the next time it will be $\frac{1}{2}$. Find three other times longer than 25 seconds for which you can forecast the height.

Exercise 2–6. If you had a stencil of the first 12 seconds in Figure 2–2, could you draw the curve for 36 seconds to 48 seconds? If the graph stopped at 25.7 seconds, could you use the stencil to continue it?

We can systematize the modular arithmetic you have been doing in Exercises 2–1 through 2–6 in formal definitions and theorems. Recall first that the **integers** are the counting numbers 1, 2, 3, . . . , their negatives −1, −2, −3, . . . , and zero.

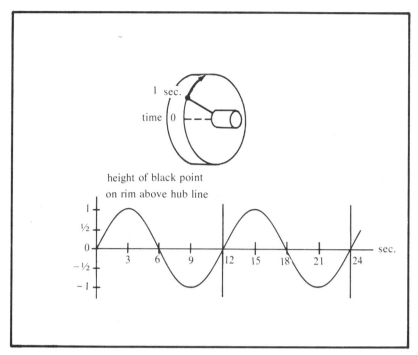

FIGURE 2–2. A PERIODIC FUNCTION.

FIGURE 2–3. PERIODICITY IN LASER LIGHT.

The light from a laser is distinguished by its near-perfect periodicity. Waves quite far apart are almost exactly alike.

Definition 2-1. Two integers a and b are **congruent modulo** an integer **m** greater than 1, if their difference is a multiple of m. In symbols, $\mathbf{a \equiv b \ (mod \ m)} \Leftrightarrow a - b = km$ for some integer k.

EXAMPLES

$5 \equiv 1 \ (mod \ 4)$, because $5 - 1 = 4 = (1)4$.

$17 \equiv 1 \ (mod \ 4)$, because $17 - 1 = 16 = (4)4$.

$18 \equiv -2 \ (mod \ 10)$, because $18 - (-2) = 20 = 2(10)$.

$18 \not\equiv -10 \ (mod \ 9)$, because $18 - (-10) = 28$, which is not a multiple of 9.

$7 \equiv 1 \ (mod \ 2, mod \ 3, and \ mod \ 6)$ but $7 \not\equiv 1$ for any other modulus m, since $7 - 1 = 6$ is a multiple of 2, 3, and 6, but of no other integers m greater than 1.

Exercise 2-7. Write 3 positive and 3 negative integers that are congruent to 2 modulo 10.

Exercise 2-8. Write 2 positive integers and 1 negative integer that are congruent to 30 modulo 360.

Exercise 2-9. Write 3 integers that are congruent to 3 modulo 12.

Exercise 2-10. Find 3 moduli m for which $-25 \equiv +25 \ (mod \ m)$.

Exercise 2-11. For what modulus m are 15 and 2 congruent? Why is there only one modulus for which they are congruent?

Exercise 2-12. Write 6 integers that are congruent to 0 modulo 5. Can you characterize all integers congruent to 0 (mod 5)?

Exercise 2-13. Write 10 integers that are congruent to 0 modulo 2. Can you characterize all integers congruent to 0 (mod 2)?

Exercise 2-14. You are programming a computer-bass for a dance band. The bass you want consists of a 4-bar phrase repeated throughout the piece. You have labeled the basic 4 measures A, B, C, and D. Which of these measures should the computer use for measure 18 of the piece? measure 15? How many measures would you expect one repetition of the song to take?

Theorem 2-1. Let a and b be integers. Then a and b are congruent modulo m if and only if $a = b + km$ for some integer k.

PROOF. The proof is based directly on Definition 2–1.

Exercise 2–15. Prove Theorem 2–1. ∎

Exercise 2–16. When is an integer a congruent to 0 modulo 10? From Theorem 2–1, a and 0 are congruent if

$$a = 0 + k\,10 = k\,10$$

for some integer k. This means that a must be a multiple $k\,10$ of 10. (Try 10, 20, -30, for instance.) When is an integer a congruent to 0 modulo 5? (Try 5, 35, -10, for instance.) Show that if a is congruent to 0 modulo 10, then it is also congruent to 0 modulo 5 ($90 = 9 \cdot 10 = 9 \cdot 2 \cdot 5$, for instance).

Exercise 2–17. Generalize Exercise 2–16 to show that if one modulus n is a multiple of another, say $n = qm$, then $a \equiv 0 \pmod{n}$ also implies $a \equiv 0 \pmod{m}$.

EQUIVALENCE CLASSES AND REPRESENTATIVE RESIDUES

Since $b, b + m, b + 2m, b - m, b - 2m, b + 3m$, and so on, are just alike modulo m, you do not need all of them for congruence arithmetic. That is why your clock does not have a number 13. Since the hours are numbered modulo 12 (24 in Europe and in military usage), you do not need a 13 o'clock; 1 o'clock suffices for 1, 13, 25, -11, and so on. You use "1" as representative for all hours that can be written $1 + k(12)$ for an integer k.

Definition 2–2. All the integers congruent to a given integer b modulo m constitute an **equivalence class** modulo m: $\{\ldots, b - 2m, b - m, b, b + m, b + 2m, \ldots\}$.

Theorem 2–2. Every integer lies in exactly one of m equivalence classes modulo m.

PROOF. Let $\bar{0}$ stand for the equivalence class

$$\bar{0} = \{\ldots, 0 - 2m, 0 - m, 0, 0 + m, 0 + 2m, 0 + 3m, \ldots\}.$$

Similarly, let

$$\bar{1} = \{\ldots, 1 - 2m, 1 - m, 1, 1 + m, 1 + 2m, 1 + 3m, \ldots\}.$$
$$\bar{2} = \{\ldots, 2 - 2m, 2 - m, 2, 2 + m, 2 + 2m, 2 + 3m, \ldots\}.$$

. . . .

$$\overline{m-1} = \{\ldots, m - 1 - 2m, m - 1 - m, m - 1,$$
$$m - 1 + m, m - 1 + 2m, m - 1 + 3m, \ldots\}.$$

Notice that you do not need a separate class $\bar{m}$, since all integers congruent to m are also congruent to zero. Now let a be any integer. Let km be the greatest multiple of m less than or equal to a. The fact that there is such a greatest multiple is a basic property of integers, called the "Archimedean property." Then $a = b + km$ and $0 \leq b < m$, since km is maximal. Then a lies in the equivalence class $\bar{b}$. There is no ambiguity about the equivalence class to which a belongs, for if a is also in the equivalence class $\bar{c}$, then a has the form $a = c + jm$, for some integer j. Set the two forms for a equal to each other and conclude that

$$c + jm = b + km.$$

Then $c = b + (k - j)m$ is in the equivalence class $\bar{b}$, and the classes $\bar{b}$ and $\bar{c}$ are the same. ∎

EXAMPLE The four equivalence classes modulo 4 are $\bar{0}$, $\bar{1}$, $\bar{2}$, and $\bar{3}$. To find which class contains -17, for instance, find the greatest multiple $k4$ of 4 for which

$$k4 \leq -17.$$

Since $-5 \cdot 4 = -20 < -17$, but $-4 \cdot 4 = -16 > -17$, $k4 = (-5)4$. Write -17 in the form

$$-17 = -5 \cdot 4 + 3.$$

Then -17 is in the class $\bar{3}$. Other members of $\bar{3}$ are -21, -13, -9, -5, -1, 3, 7, 11, 15, and 119.

Exercise 2–18. To what equivalence class does 38 belong modulo 4?

FIGURE 2–4. MUSICAL SCALES AS EXAMPLES OF MODULAR SYSTEMS.

The C-major scale, C, D, E, F, G, A, B, C, is a mod 7 system. The chromatic scale, C, C#, D, D#, E, F, F#, G, G#, A, A#, B, C, is a mod 12 system.

Exercise 2–19. Find one member of each equivalence class modulo 5.

Exercise 2–20. How many different equivalence classes are there modulo 2? Characterize their respective members. See Figure 2–5.

FIGURE 2–5. SYSTEMS HAVING PERIOD 2.

Modulus 2 arithmetic describes changes that undo their effects when performed a second time.

Exercise 2–21. Arrange the following integers in equivalence classes modulo 5:

16	127	−5670	0	−17	14	33
99	21	−123	1111	71	45	32
29	11	4	−3	−7	12	−27
19	−8	34	76	92	75	105

The equivalence classes of integers have their own arithmetic. Two equivalence classes can be added or multiplied by adding or multiplying their representatives. In Theorem 2–3 we show that one representative from each class is enough to determine the sum and product classes.

Theorem 2–3. Let a be congruent to r modulo m and let b be congruent to s modulo m. Then

$$a + b \equiv r + s \quad \text{and} \quad a \cdot b \equiv r \cdot s \;(\text{mod } m).$$

PROOF. See Exercises 2–22 and 2–23.

Exercise 2–22. Express a as $r + jm$ and b as $s + km$. Then compute $a + b$ and show that it is congruent to $r + s \;(\text{mod } m)$.

Exercise 2–23. Express a as $r + jm$ and b as $s + km$. Then compute $a \cdot b$ and show that it is congruent to $r \cdot s \;(\text{mod } m)$. ∎

From Theorem 2–3, it makes no difference which representative of an equivalence class is used in congruence arithmetic, so the usual policy is to use a small one!

The arithmetic here parallels that of Chapter 5, and both developments illustrate a typical mathematical approach: First, the objects are separated into equivalence classes (in Chapter 5 this amounts to telling which matrices are equal). Then arithmetic operations are defined on these classes. Although the operations are called addition and multiplication both for matrices and for integers modulo m, you will notice that the familiar names stand for quite different operations, each appropriate to its system.

Definition 2–3. A **representative** of an equivalence class modulo m is a member of the class. The **least non-negative residues** modulo m are $0, 1, 2, \ldots, m - 1$. A **complete set of residues** modulo m is any set of m integers, one from each equivalence class.

EXAMPLE   The least non-negative residues modulo 4 are 0, 1, 2, and 3. Their arithmetic modulo 4 represents the arithmetic of all the integers modulo 4 (by Theorem 2–3):

+	0	1	2	3
0	0	1	2	3
1	1	2	3	0
2	2	3	0	1
3	3	0	1	2

·	0	1	2	3
0	0	0	0	0
1	0	1	2	3
2	0	2	0	2
3	0	3	2	1

Why does the addition table give "1" opposite 2 and 3, showing that $2 + 3$ is 1? From the definition of congruence, $2 + 3 = 5$ is congruent to 1 modulo 4.

Notice that here $2 + 2 \equiv 0$. Perhaps one thing you were sure of in mathematics before you read this book was that "2 plus 2 makes 4"! Suppose you had to explain to a friend that sometimes 2 plus 2 can equal zero. Could you make it seem plausible? It would be easiest to rely on something cyclic of period four, such as the four seasons of the year or the four quarters of the moon, or 25-cent pieces as quarters of a dollar. It should be fairly easy to convince your friend that if an amount of money is represented by dollar bills plus two quarters and a second amount of money, also dollar bills plus two quarters, is added to it, then the resulting sum of money is a whole number of dollars with no coins necessary.

Exercise 2–24. Write the least non-negative residues modulo 5. Write their addition and multiplication tables.

Exercise 2–25. Write the least non-negative residues modulo 3. Show that $\{-1, 0, 1\}$ is also a complete set of residues, and, using this latter set, write the addition and multiplication tables modulo 3.

Exercise 2–26. Use the multiplication table from Exercise 2–24 to write the first 10 powers of 2 modulo 5: $2^1 \equiv 2$, $2^2 = 2 \cdot 2 \equiv 4$, $2^3 = 2^2 \cdot 2 \equiv 4 \cdot 2 \equiv 3$, and so on, through $2^{10} = 2^9 \cdot 2$. Observe that the powers of 2 (mod 5) are periodic.

Exercise 2–27. Write the multiplication table of a complete set of residues modulo 6. You may choose to use some negatives, such as $-1 \equiv 5$.

Exercise 2–28. Write the multiplication table of a complete set of residues modulo 7. By using negatives you can avoid integers more than 4 units from zero.

Exercise 2–29. Notice that modulo 4 the product $2 \cdot 2$ is zero, although neither factor is zero. What is $2 \cdot 3$ modulo 6? Is there any product $a \cdot b$ with neither factor congruent to zero but $a \cdot b \equiv 0$ modulo 3? modulo 5? modulo 7? Find $3 \cdot 6$ modulo 9. Can you form a conjecture about moduli having this feature? Can you support your conjecture with further examples? with proof?

Exercise 2–30. Work these arithmetic problems modulo 9:
$$129(132 + 5763) \equiv ?$$
$$2^4(3 + 12 + 15 + 1) \equiv ?$$
$$10 + 12 + 16 + 21 \equiv ?$$

Exercise 2–31. In official time announcements the minutes are usually given in the form of least non-negative residues modulo 60. What other complete set of residues is in common use? (For instance, how do you sometimes say "10:55"?)

Exercise 2–32. How is the congruence $32 \equiv -68 \pmod{100}$ used in making change? What other congruences are used?

DIVISORS

Definition 2–4. We have used the fact that an integer m is a **multiple** of an integer d if there is an integer k for which $m = kd$. If d is a non-zero integer, then d is a **divisor** of an integer m if there is a quotient integer q for which $m = qd$. We write "d divides m" in symbols as "**d | m.**"

EXAMPLES -21, 309, 123, and 0 are multiples of 3.
0 is a multiple of 1, 2, 7, 769, -124, and 0.
2 is a divisor of -18, 16, 14, -2, 0, 4, and 6.
0 cannot be tested as a divisor, since only non-zero integers d enter into the definition.

Exercise 2–33. Find all the positive and negative divisors of 12. Is zero a divisor of 12? Is zero a multiple of 12?

Exercise 2–34. Find all the positive and negative divisors of 18. Is zero a divisor of 18? Is zero a multiple of 18?

Exercise 2–35. How can you use the positive divisors of an integer to find the negative divisors? Prove directly from Definition 2–2 that if $+d$ is a divisor of m, then $-d$ is also a divisor of m.

Exercise 2–36. Can an integer have divisors greater than itself? (Consider negative integers.) Can a positive integer have a positive divisor greater than itself?

Theorem 2–4. Let d be an integer greater than 1. Then d is a divisor of an integer a if and only if a is congruent to 0 modulo d.

PROOF.   See Exercises 2–37 and 2–38.

Exercise 2–37. Suppose that an integer $d > 1$ divides an integer a. Use Definition 2–4 to express a in terms of d. Then interpret this as a congruence modulo d according to Definition 2–1.

Exercise 2-38. Let $d > 1$ be an integer and let $a \equiv 0 \pmod{d}$. Use Definition 2-1 and then Theorem 2-1 to show that d divides a. ∎

You have learned from experience that integers divisible by 10 end in **EXAMPLE** a zero: 100, 20, -30, 5280, and so on. By Theorem 2-4, the integers divisible by 10 are those that are congruent to zero modulo 10. Write the integer 5280 in powers of 10: The 5 is in the thousands place, and thus stands for $5 \cdot 10^3$. The 2 is in the hundreds place, and contributes $2 \cdot 10^2$. The 8 in the tens place contributes $8 \cdot 10$, and the 0 in the units place contributes no units. Then $5280 = 5 \cdot 10^3 + 2 \cdot 10^2 + 8 \cdot 10 + 0 \cdot 1$ is congruent to $5 \cdot 0^3 + 2 \cdot 0^2 + 8 \cdot 0 + 0 \cdot 1 \equiv 0 \pmod{10}$. Notice that in reducing the sum one term at a time modulo 10 you rely on Theorem 2-3.

Exercise 2-39. You know that the integers that are divisible by 5 are the ones that end in 0 or in 5. Show how to express an integer as the sum of its digits multiplied by the appropriate power of 10 and reduce the result modulo 5. Prove that the 0-or-5 test works.

Exercise 2-40. You have noticed that even integers end in 0, 2, 4, 6, or 8. Prove that there is a solid basis for this observation. In particular, show by writing 32141 in powers of 10 and reducing modulo 2 that 32141 is not divisible by 2 (not even).

Exercise 2-41. Prove that a test for divisibility by 4 cannot depend on the last digit alone, by finding a number n ending in an even digit with n divisible by 4, and another number m ending in the same digit but not divisible by 4.

Exercise 2-42. Prove by consideration of powers of 10 that an integer is divisible by 4 if and only if its last two digits form an integer that is divisible by 4.

Exercise 2-43. Develop a test for divisibility by 8. (The previous two exercises provide a hint.)

Many number tricks depend on the idiosyncrasies of the number 9 in our 10-base system. Perhaps you learned (or wished you could remember) a trick for multiplying an integer by 9. You decreased the integer, say 7, by 1, and used the result, 6, as the tens digit. Then you selected the units digit, 3, that adds to 6 to give 9: $7 \cdot 9 = 63$. Or you may have held your ten fingers in a row, depressed the seventh one, and counted 6 to the left, 3 remaining to the right, a physical realization of the first method. Both of these tricks depend on the fact that $7 \cdot 9 = (6 + 1)9 = 6 \cdot 9 + 1 \cdot 9 = 6(10 - 1) + 9 = 60 + (9 - 6)$.

"Casting out nines" is a very old trick that was of some practical advantage to the human adders who preceded the now-ubiquitous desk computer. Suppose a bookkeeper had summed a column of prices

$$\begin{array}{r} \$176.29 \\ 39.95 \\ 107.68 \\ 55.00 \\ 34.07 \\ 29.95 \\ \underline{15.54} \\ \hline \$458.48 \end{array}$$

Then as a check he would add together the digits in each price, reducing the amount modulo 9:

$176.29 $\quad 1 + 7 + 6 + 2 + 9 = 25 \equiv 7 \ (\text{mod } 9)$

$\ 39.95 $\quad 3 + 9 + 9 + 5 \equiv 3 + 0 + 0 + 5 \equiv 8$

$107.68 $\quad 1 + 0 + 7 + 6 + 8 \equiv 1 + 0 - 2 + 6 - 1 \equiv 4$, and so on.

Then he would add these amounts modulo 9 to be sure they gave the same result as the total price modulo 9:

$$7 + 8 + 4 + 1 + 5 + 7 + 6 \equiv 7 - 1 + 4 + 1 - 4 - 2 - 3 \equiv 2$$
$$\$458.48 \quad 4 + 5 + 8 + 4 + 8 \equiv (4 + 5) - 1 + 4 - 1 \equiv 2$$

The following theorem shows why casting out nines works.

> **Theorem 2–5.** An integer written in decimal (base 10) notation is congruent modulo 9 to the sum of its digits.

PROOF. The integer

$$a = a_n a_{n-1}, \ldots, a_2 a_1 a_0$$

in decimal notation means

$$a = a_n 10^n + a_{n-1} 10^{n-1} + \cdots + a_2 10^2 + a_1 10^1 + a_0 10^0,$$

with the convention that $10^0 = 1$, because of the meaning attached to the position of each digit.

Modulo 9, $10 \equiv 1$, so $10^i \equiv 1$ for each positive degree i. Then

$$a \equiv a_n 1 + a_{n-1} 1 + \cdots + a_2 1 + a_1 1 + a_0 1,$$

that is, the sum of the digits of a. ∎

By Theorem 2–3 you can substitute congruent quantities for each other in modular arithmetic without changing the results, so adding sums of digits should produce a sum congruent to the sum of a column of figures.

Exercise 2–44. Find mistakes among the following additions by casting out nines.

34	438	189	23
28	392	276	72
90	87	12	61
127	888	1102	37
31	256	103	39
300	2052	1582	232

Exercise 2–45. One of the additions in Exercise 2–44 is incorrect, even though it checks modulo 9. Find which one it is and explain why casting out nines did not expose it.

Exercise 2–46. How many different equivalence classes are there modulo 9? Then there would be one chance in __ of getting the right residue when casting out nines by chance alone.

Exercise 2–47. Find the least non-negative residue of 10 modulo 3. Follow the scheme of Theorem 2–5 to develop a test for divisibility of integers by 3.

Exercise 2–48. Test the following integers for divisibility by 2, 3, 4, 5, 9, 10, and 100:

170	1100	1978	111
235	1964	134	333
945	1966	450	72
123	1974	225	137
126	1976	101	1331

Definition 2–5. A **common divisor** of two integers is an integer that divides both of them. Let a and b be two integers not both zero. Then their **greatest common divisor, (a, b),** is a common divisor that is divisible by every common divisor. Usually we restrict attention to the positive greatest common divisor.

EXAMPLE

Let $a = 189$, $b = 168$. Factor them into irreducible factors: $a = 3^3 \cdot 7$, $b = 2^3 \cdot 3 \cdot 7$. Notice that 1, 3, 7, and 21 are common divisors of a and b, and so are their negatives -1, -3, -7, and -21. Their positive greatest common divisor is $(189, 168) = 21$, for 21 divides both 189 and 168 and is divisible by all the common divisors.

In general, to find the greatest common divisor of factored integers, use the smaller (minimal, or "min") power of each prime in the factorizations. In the example, 2 does not divide 189, or you can say 2 appears to the power zero. With this convention, you find

$$(a, b) = 2^{\min(0,3)} \cdot 3^{\min(3,1)} \cdot 7^{\min(1,1)}$$

$$= 2^0 \cdot 3^1 \cdot 7^1 = 21.$$

Exercise 2–49. Find (100, 12).

Exercise 2–50. Find (48, 198).

Exercise 2–51. Find (1980, 1701). (Recall that $3^0 = 1$, and $5^0 = 1$ by convention.)

For solving congruences we need a way to find (a, m) in terms of a and m, as

$$(a, m) = ra + sm,$$

where r and s are integers. There is an "algorithm," or stepwise procedure that will always give r and s, and, incidentally, (a, m) as well.

> **Definition 2–6.** The Euclidean algorithm is a stepwise process for finding $(a, m) = ra + sm$. It is explained here and illustrated for the case $a = 132$, $m = 600$.

First, divide m by a, getting a remainder r_1 less than a:

$$600 = 132 \cdot 4 + 72, \quad \text{with} \quad r_1 = 72 < 132$$

Then, divide a by the first remainder r_1, getting a second remainder r_2 less than r_1:

$$132 = 72 \cdot 1 + 60, \quad \text{with} \quad r_2 = 60 < 72$$

Continue dividing each previous remainder by the new smaller remainder until a division comes out even (zero remainder):

$$72 = 60 \cdot 1 + 12, \quad \text{with} \quad r_3 = 12 < 60$$

$$60 = 12 \cdot 5 + 0$$

(a, m) is the last non-zero remainder. To find r and s, use the division steps in reverse order:

$$72 = 60 \cdot 1 + 12, \text{ so that}$$

$$12 = 72 - 60 \cdot 1 \tag{1}$$

From the previous division,

$$132 = 72 \cdot 1 + 60, \text{ so that}$$

$$60 = 132 - 72 \cdot 1.$$

Use this formula for 60 in (1), to write

$$12 = 72 - 60 \cdot 1 = 72 - (132 - 72 \cdot 1)$$

$$= 72(1 + 1) - 132$$

$$12 = 72 \cdot 2 - 132 \tag{2}$$

From the first division,

$$600 = 132 \cdot 4 + 72, \text{ so that}$$
$$72 = 600 - 132 \cdot 4.$$

Use this formula for 72 in (2), to write

$$12 = 72 \cdot 2 - 132 = (600 - 132 \cdot 4)(2) - 132$$
$$= (-9)(132) + (2)(600).$$

Then r is -9 and s is 2. As a check, calculate

$$(-9)(132) + (2)(600) = -1188 + 1200 = 12 = (132, 600).$$

Find $(210, 495)$ in terms of 210 and 495.   **EXAMPLE**

$$495 = 210 \cdot 2 + 75, \quad \text{with} \quad 75 < 210$$
$$210 = 75 \cdot 2 + 60, \quad \text{with} \quad 60 < 75$$
$$75 = 60 \cdot 1 + 15, \quad \text{with} \quad 15 < 60$$
$$60 = 15 \cdot 4 + 0$$

$$(210, 495) = 15$$

$$75 = 60 \cdot 1 + 15 \rightarrow 15 = 75 - 60 \cdot 1$$
$$210 = 75 \cdot 2 + 60 \rightarrow 60 = 210 - 75 \cdot 2$$
$$15 = 75 - (210 - 75 \cdot 2) \cdot 1 = -210 + 3 \cdot 75$$
$$495 = 210 \cdot 2 + 75 \rightarrow 75 = 495 - 210 \cdot 2$$
$$15 = -210 + 3(495 - 210 \cdot 2) = (-7)(210) + (3)(495).$$

$$r = -7, s = 3.$$

Theorem 2-6. Let a and m be positive integers. Then there are integers r and s for which

$$(a, m) = ra + sm.$$

PROOF. The Euclidean algorithm always yields r and s: Each division results in a remainder less than the divisor. Therefore each new remainder is less than the one before it, so that eventually some remainder must be zero (a decreasing sequence of non-negative integers). Substitution from the division steps in reverse order produces a formula for the last non-zero remainder in the form $ra + sm$.

To show that the last non-zero remainder is indeed (a, m), first note that from the form $ra + sm$ it is divisible by every common divisor of a and m. For instance, in the example above,

$$15 = (-7)(210) + (3)(495)$$

is divisible by the common divisor 5 of 210 and 495, since it can be written as

$$15 = [(-7)(42) + (3)(99)](5).$$

Also, the last non-zero remainder is a common divisor of all the remainders and of a and m. For instance, in the example above, the last division,

$$60 = 15 \cdot 4 + 0,$$

shows that 15 divides 60. Then in the previous step

$$75 = 60 \cdot 1 + 15 = 4 \cdot 15 + 1 \cdot 15,$$

so 15 divides 75. Then from $210 = 75 \cdot 2 + 60 = 10 \cdot 15 + 4 \cdot 15$, 15 divides 495. ∎

Exercise 2–52. Find $(18, 66)$ by the Euclidean algorithm in terms of 18 and 66.

In their book *Discovering Number Theory* (W. B. Saunders Company, 1972, pages 14–16), the authors show a pictorial version of the Euclidean algorithm based on length measurements. In this physical interpretation the algorithm suggests itself quite naturally.

LINEAR CONGRUENCES

I have just bought theater tickets for a party of friends, at $2.75 per ticket. I paid in bills, that is, denominations of one or more dollars, with no coins, and received in change exactly 75 cents in coins. How many tickets did I buy?

Suppose I bought x tickets. Then the total price was x times $2.75, and from the 75 cents in change,

$$\$2.75x \equiv \$0.25 \text{ modulo } \$1$$

or $3x \equiv 1$ modulo 4, counting in quarters.

(You can drop the $2, since it is only the change less than a dollar that you have information about, anyway.)

This congruence is easy to solve by trial and error, since there are only 4 distinct equivalence classes modulo 4.

$$3 \cdot 0 \equiv 0 \not\equiv 1$$
$$3 \cdot 1 \equiv 3 \not\equiv 1$$
$$3 \cdot 2 \equiv 2 \not\equiv 1$$
$$3 \cdot 3 \equiv 1$$

Then the number of tickets was congruent to 3 modulo 4. It may have been 3 tickets, or 7 tickets, or 11 tickets, or more. With more data, such as what bills I paid in, how stingy I am, and so on, it might be possible to infer more about the number of tickets.

Exercise 2–53. Originally, when planning my theater party, I issued seven invitations. How many did not accept my invitation? (Remember that I plan to attend.) This exercise has two answers that are correct mathematically. Which do you consider more probable from your social experience?

Exercise 2–54. You are to administer medicine to a sick pet every 4 hours. If you start at 2 o'clock, show that you will not be dosing your pet at midnight, even if the regimen continues indefinitely.

You have driven from Suburbia to Metroburg, back to Suburbia, then **EXAMPLE** back to Metroburg. The three trips have increased the units dial on your odometer by 1 mile, but you failed to note the tens dial setting when you started. How many miles is Suburbia from Metroburg?

Suppose Suburbia is x miles from Metroburg. Then $3x = 1$, or 11, or 21, or 31, and so on. The only information you have is about the units digit, that is, about x modulo 10. You need solutions to the linear congruence

$$3x \equiv 1 \text{ modulo } 10.$$

By trial and error,

$3 \cdot 0 \equiv 0$	$3 \cdot 3 \equiv 9$	$3 \cdot 6 \equiv 8$
$3 \cdot 1 \equiv 3$	$3 \cdot 4 \equiv 2$	$3 \cdot 7 \equiv 1$
$3 \cdot 2 \equiv 6$	$3 \cdot 5 \equiv 5$	$3 \cdot 8 \equiv 4$
		$3 \cdot 9 \equiv 7$

There is just one solution modulo 10, $x \equiv 7$. Then possible distances between Suburbia and Metroburg are 7 miles, 17 miles, 27 miles, or more generally $7 + 10k$ miles, where k is an integer. Of course, you might well be able to estimate the distance to the nearest 10 miles and so determine it exactly.

Suppose you have made an x-mile trip (x an integer) 6 times. Then **EXAMPLE** the resulting change in the units digit of your odometer cannot be 3. The change would be

$$6 \cdot 0 \equiv 0, \quad \text{if} \quad x = 0,$$
$$6 \cdot 1 \equiv 6, \quad \text{if} \quad x = 1,$$
$$6 \cdot 2 \equiv 2, \quad \text{if} \quad x = 2,$$
$$6 \cdot 3 \equiv 8, \quad \text{if} \quad x = 3,$$
$$6 \cdot 4 \equiv 4, \quad \text{if} \quad x = 4, \text{ and so on.}$$

The linear congruence $6x \equiv 3$ modulo 10 has no solutions for x.

You can analyze this example to see what it is about the congruence $6x \equiv 3$ (mod 10) that keeps it from having solutions. For each trial x, in your search for a solution, you form $6x$ and then subtract as many 10's,

$k \cdot 10$, as you can without making the result negative. For instance,

$$6 \cdot 9 = 54 \equiv 54 - 5 \cdot 10 \equiv 4 \,(\mathrm{mod}\ 10).$$

In each case $6x - k10$ does not yield 3, because $6x - k10 = 2(3x - k5)$ *is* divisible by 2, and 3 is *not* divisible by 2.

You can use this analysis to suggest a general theorem.

Theorem 2–7. Let d be a common divisor of a and m and $d \nmid b$. Then the congruence

$$ax \equiv b \,(\mathrm{mod}\ m)$$

has no solutions.

PROOF. From Theorem 2–1 all the integers congruent to ax modulo m have the form

$$ax + km, \text{ where } k \text{ is an integer.}$$

If d is a common divisor of a and m, then $a = dq$ for some quotient q and $m = dn$ for some quotient n. Then $ax + km = dqx + kdn = d(qx + kn)$ is divisible by d for every value of x and so cannot be b if d does not divide b. ∎

Exercise 2–55. Express the problem of Exercise 2–54 as a congruence problem

$$4k + 2 \equiv 0 \,(\mathrm{mod}\ 12),$$

transpose the 2, and apply Theorem 2–7.

Exercise 2–56. Prove that the hypothesis of Theorem 2–7 is met, that is, there is a common divisor d of a and m that fails to divide b, if and only if the greatest common divisor (a, m) fails to divide b. Then Theorem 2–7 could be stated equivalently: If $(a, m) \nmid b$, then $ax \equiv b \,(\mathrm{mod}\ m)$ has no solution.

EXAMPLE   You want to buy some media time for several broadcasts of a taped political message that runs 25 seconds, plus one broadcast of a 160-second tape on election eve. The time, however, is sold in multiples of 30 seconds. How much time will you buy?

If you buy x 30-second quantities of time, you will have $30x$ seconds of broadcast. If your 25-second message is repeated exactly k times, and your 160-second tape played once, then

$$30x = k25 + 160 \tag{1}$$

or in congruence form, by Theorem 2–1,

$$30x \equiv 160 \,(\mathrm{mod}\ 25).$$

Since $(30, 25) = 5$ divides 160, you can divide equation (1) by 5:

$$6x = k5 + 32,$$

or in congruence notation, by Theorem 2–1,

$$6x \equiv 32 \ (\text{mod } 5),$$

$$\text{or } 1x \equiv 2 \ (\text{mod } 5).$$

The basic solution (mod 5) is $x = 2$, but this hardly satisfies your advertising needs, since it would call for repeating your 25-second message *minus* four times! The other solutions (mod 25) are $2 + 5$, $2 + 2 \cdot 5$, $2 + 3 \cdot 5$, and $2 + 4 \cdot 5$. Every solution is congruent to one of them (mod 25). The smallest nontrivial solution is $x = 7$, calling for buying 7 30-second quantities of time, or 210 seconds, which will be enough to allow $(210 - 160)/25 = 2$ repetitions of your shorter message. Figure 2–6 illustrates the periodic recurrence of the solution.

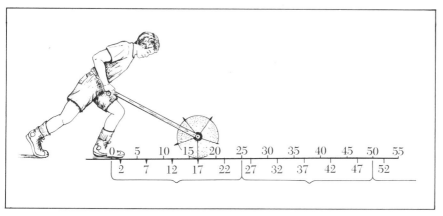

FIGURE 2–6.

Exercise 2–57. Suppose you want at least 10 repetitions. How much time should you buy?

Exercise 2–58. Suppose you want at least 100 repetitions. How much time should you buy?

This example and others have shown you that some linear congruences *do* have solutions. In fact, Theorem 2–8 will show that the linear congruences described in Theorem 2–7 are the only ones that do *not* have solutions.

Theorem 2–8. Let $g = (a, m)$, and let g divide b. Let $a = ga'$, $m = gm'$, $b = gb'$. Then the linear congruence $ax \equiv b \pmod{m}$ has g solutions modulo m, a basic solution x_0 of the congruence $a'x \equiv b' \pmod{m'}$, and the $g - 1$ related solutions $x_0 + m'$, $x_0 + 2m'$, $\ldots$, $x_0 + (g - 1)m'$.

PROOF. If $ax \equiv b \pmod{m}$, then by Theorem 2–1 $ax = b + km$, for some integer k. Then

$$ga'x = gb' + kgm',$$

from which

$$a'x = b' + km'. \tag{1}$$

From Theorem 2–6 there are integers r and s for which

$$g = (a, m) = ra + sm, \text{ or, dividing by } g,$$

$$1 = ra' + sm'. \tag{2}$$

Transpose terms in (2) to show the similarity to (1):

$$a'r = 1 + (-s)m'.$$

Multiply by b':

$$a'(rb') = b' + (-sb')m'.$$

Now you see that rb' is a solution x_0 for x in (1), with $k = -sb'$.

Exercise 2–59. You have one solution $x_0 = rb'$, for which $a'x_0 \equiv b' \pmod{m'}$, with equation (1) expressed in congruence form. Show that if y is another solution of the congruence, so that $a'y \equiv b' \pmod{m'}$, then $y \equiv x_0 \pmod{m'}$. Suggestion: From (2) you can write a congruence

$$ra' \equiv 1 \pmod{m'},$$

so that you can find y from $a'y$ as $r(a'y) \equiv (ra')y \equiv 1 \cdot y \equiv y \pmod{m'}$.

Exercise 2–60. Show that since $x_0 \equiv rb'$ is a solution for equation (1), it is a solution for the congruence $ax \equiv b \pmod{m}$.

Exercise 2–61. Show that if x_0 is a solution for $ax \equiv b \pmod{m}$, then so is $x_0 + im'$, where i is an integer. ∎

Exercise 2–62. Summarize previous results to prove: The linear congruence $ax \equiv b \pmod{m}$ has solutions if and only if $(a, m) \mid b$.

In Exercises 2–63 through 2–82, some of the linear congruences have no solution because a and m have a common divisor that does not divide b. Find what the divisor is, and cite Theorem 2–7 to prove that there is no solution. In case $(a, m) \mid b$ for a congruence, find the basic solution modulo m' and all the other solutions modulo m.

Exercise 2–63. $21x \equiv 14 \pmod{28}$.

Exercise 2–64. $9x \equiv 8 \pmod{12}$.

Exercise 2–65. $6x \equiv 8 \pmod{14}$.

Exercise 2–66. $2x \equiv 6 \pmod{4}$.

Exercise 2–67. $2x \equiv 1 \pmod{4}$.

Exercise 2–68. $8x \equiv 12 \pmod{24}$.

Exercise 2–69. $8x \equiv 4 \pmod{10}$.

Exercise 2–70. $2x \equiv 1 \pmod{5}$.

Exercise 2–71. $3x \equiv 2 \pmod{7}$.

Exercise 2–72. $3x \equiv -9 \pmod{6}$.

Exercise 2–73. $10x \equiv -12 \pmod{18}$.

Exercise 2–74. $2x \equiv -3 \pmod{6}$.

Exercise 2–75. $3x \equiv 0 \pmod{2}$.

Exercise 2–76. $3x \equiv 0 \pmod{3}$.

Exercise 2–77. $3x \equiv 0 \pmod{4}$.

Exercise 2–78. $28x - 2 \equiv 0 \pmod{7}$.

Exercise 2–79. $9x + 21 \equiv 0 \pmod{6}$.

Exercise 2–80. $8x + 8 \equiv 9 \pmod{11}$.

Exercise 2–81. $12x \equiv 20 \pmod{8}$.

Exercise 2–82. $2x \equiv 5 \pmod{13}$.

SOME FEATURES OF CONGRUENCE ARITHMETIC

As you have seen, modular arithmetic, especially for a small modulus m, is easy, because there are only m different numbers in the system. For this reason, modular arithmetic serves as a miniature showcase for features of interest in arithmetics.

Unique Factorization

You can write 6 in factored form

$$6 = 2 \cdot 3.$$

If you are working modulo 4, you have

$$6 = 2 \cdot 3 \text{ and also, } 6 \equiv 1 \cdot 2, \text{ or } 2.$$

Here are two different factorizations for the same integer; even the number of factors is different.

Here then is a property conspicuous by its absence in modular arithmetic. The lack here may make you conscious of the unique factorization in many mathematical systems, such as the integers, polynomials with integer coefficients, and so on.

Proper Divisors of Zero

Modulo 12, 6 is not congruent to zero and 10 is also not congruent to zero, yet their product *is* congruent to zero:

$$6 \not\equiv 0, \ 10 \not\equiv 0, \text{ but } 6 \cdot 10 = 60 \equiv 0 \text{ (mod 12)}.$$

In this case 6 and 10 are called **proper divisors of zero,** because neither is zero (mod 12) but their product is congruent to zero.

Here again is a feature not found in the arithmetic of integers or of fractions. In the case of integers or fractions, if you are told that a product ab is zero, then you can deduce that one of the factors at least is zero. Sometimes you use this fact in solving equations. Suppose you are looking for integers x that satisfy the equation

$$x^2 - 10x + 21 = 0.$$

You can write the left side as a product

$$(x - 7)(x - 3) = 0.$$

Then you say that either $x - 7$ must be zero or $x - 3$ must be zero if their product is to yield zero.

Now take not an equation but the *congruence*

$$x^2 - 10x + 21 \equiv 0 \text{ modulo 12}.$$

There are two correct factorizations, among others:

$$(x - 7)(x - 3) \equiv 0 \quad \text{and} \quad (x - 1)(x + 3) \equiv 0$$

and the congruence has more distinct solutions (mod 12) than its degree, four in each interval of length 12: $x \equiv 1, 3, 7, $ and 9.

Additive and Multiplicative Identities

The 0 equivalence class $\ldots, -2m, -m, 0, m, 2m, \ldots$ plays the role in modular arithmetic that 0 plays among the integers or the fractions. It can be added to any number without changing it (mod m).

Similarly, 1, and all integers $1 + km$, have the multiplicative identity property: They can multiply any number without changing it.

For comparison, the even integers with ordinary integer multiplication supply an example of a system with no multiplicative identity:

$$\ldots, -6, -4, -2, 0, 2, 4, 6, \ldots.$$

Multiplicative Inverses

If a mathematical system has a multiplicative identity 1, then a^{-1} is a multiplicative inverse of a if $a(a^{-1}) = (a^{-1})a = 1$. In the system of fractions any non-zero fraction n/d has a multiplicative inverse d/n, since $(n/d)(d/n) = nd/dn = 1$ and $(d/n)(n/d) = dn/nd = 1$.

In the mathematical system of integers modulo 4, the table

×	0	1	2	3 (mod 4)
0	0	0	0	0
1	0	1	2	3
2	0	2	0	2
3	0	3	2	1

gives all the multiplicative facts at a glance. Modulo 4, two of the nonzero integers, 1 and 3, have inverses:

$1^{-1} \equiv 1$, because $1 \cdot 1 \equiv 1$, and $3^{-1} \equiv 3$, because $3 \cdot 3 \equiv 1$.

However, there is no integer x for which $2x \equiv 1 \pmod 4$, even though 2 is not zero (mod 4).

To find a multiplicative inverse for an integer a modulo m, you need a solution for the congruence

$$ax \equiv 1 \pmod m.$$

From Exercise 2–62, then, a has an inverse if and only if $(a, m) \mid 1$, that is, if and only if $(a, m) = 1$.

Exercise 2–83. Find $(2, 4)$ and use the previous sentence to show why 2 has no multiplicative inverse (mod 4).

Exercise 2–84. Suppose $(a, m) = 1$. How many different multiplicative inverses does a have modulo m?

Congruence arithmetic is included in the branch of mathematics called "number theory." You might follow up interest in this subject by reading *Discovering Number Theory*, J. E. Maxfield and M. W. Maxfield, W. B. Saunders Company, 1972.

REVIEW OF CHAPTER 2

1. Define $a \equiv b$ modulo m. Give examples of integers that are congruent for one modulus but are not congruent for another modulus.

2. Write two members, one of them negative, of each equivalence class modulo 6.

3. Write the least non-negative residues modulo 6 and calculate their addition and multiplication tables modulo 6.

4. Write the integer 4365 in powers of 10 and reduce modulo 3. Determine whether 3 is a divisor of 4365.

5. How can you determine without actually dividing whether an integer is divisible by 2? 5? 10? 100? 25? 4? 3? 9?

6. Write a linear congruence that has no solution.

7. Write a linear congruence that has 2 solutions modulo 14.

8. Use the Euclidean algorithm to find (486, 512) in terms of 486 and 512.

9. Congruence arithmetic modulo 10 does not have **unique factorization**, and there are **proper divisors of zero** modulo 10. What do these terms mean? Illustrate with examples, if possible.

10. There are integers having no **multiplicative inverses** modulo 10. Explain what the term means, and give an integer with an inverse and one with no inverse.

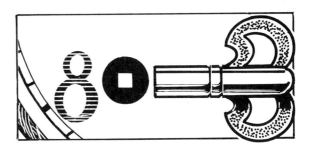

CHAPTER 2 AS A KEY TO MATHEMATICS

Congruence arithmetic is naturally adapted to applications exhibiting periodicity, but you have been coping with those successfully without a special arithmetic. What is the advantage of studying congruences, then? One of its major advantages is that you already understand it from experience. You know that multiples of 50 cents will not produce an amount ending in 75 cents, and without stating it in formal terms, you have a good idea why. This position is perfect for learning to prove theorems for yourself. It is pleasant to discover features of modular arithmetic for yourself and to be able to state and prove theorems that describe your discoveries.

Modular arithmetic is also valuable in its comparisons and in its contrasts with other arithmetics, helping you concentrate on the features of ordinary arithmetic from a fresh point of view.

Perhaps most important of all is realizing that *arithmetic* can be used in the plural; that there can be different arithmetics for different purposes, each with its own rules and interior coherence.

CHAPTER 3

"Contrariwise," continued Tweedledee, "if it was so, it might be; and if it were so, it would be; but as it isn't, it ain't. That's logic."

Through the Looking-Glass, *by Lewis Carroll*

Mathematical proofs are "deductive." They move according to formal laws from each established point to the next. Within the prescribed areas where such formal reasoning applies, the laws of logic can be used like those of arithmetic or algebra, and can even be turned over to a computer.

DEDUCTIVE LOGIC

Rules for Reasoning

At Oak Ridge National Laboratory in Tennessee, many disgruntled, frustrated scientists were late reporting for work one day. It turned out that a new security guard had misinterpreted his instructions. He had been told to guard Area B by inspecting each employee's pass and admitting only those whose passes showed a letter "B." The difficulty arose because most passes listed several letters, and the new guard would not admit anyone whose pass had a C, a G, *and* a B, for instance. Of course, the intended interpretation was that such an employee should be admitted to Area C *and* he should be admitted to Area G *and* he should be admitted to Area B.

In this example suppose *B* stands for "This employee may enter Area B," and *B-and-C* stands for "This employee may enter Area B *and* this employee may enter Area C." Then it is common to say that *B* can be *deduced* from *B-and-C*, or that *B follows from B-*and*-C*, or that *B is implied by B-*and*-C*. All these are ways to express our feeling that *B* is somehow present within *B-and-C*, so that without any further information *B-and-C* tells us *B*. "Implies" means literally "enfolds"; *B-and-C* implies *B* in the sense that it holds *B* folded within itself.

Classical statement logic provides a formal system restricted enough so that we can check whether a conclusion is correctly deduced from the premises. Certain patterns of deduction occur so often that they have names of their own. They are recognizable as patterns regardless of the meanings of the statements involved. A similar case occurs in law court, where the pattern of an "alibi" defense is recognizable regardless of the content of the alibi or the offense: The defense tries to establish that the defendant was elsewhere when the offense was committed and asks the jury to conclude from this that the defendant is not guilty of the offense.

STATEMENTS IN A SPACE

Formal deductive reasoning is severely restricted, so that definite ground rules can be written for it. In fact, you can put it all in a box! The first step in a deductive argument is to agree on a **universe of discourse,** or **space,** U. It can be suggested schematically by a rectangular box, as in Figure 3–1. The box stands for all the objects that could reasonably enter the discussion; for instance, for one problem, U might be all people now living, for another all giraffes, for another all whole numbers, and so on. In a guessing game like Twenty Questions, the first questions try to set the universe of discourse: as animal, vegetable, or mineral, as fact or fiction, as living or dead, and so on.

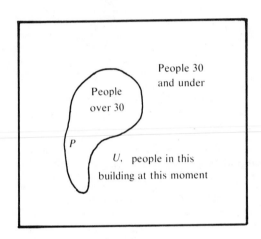

FIGURE 3–1. A STATEMENT IN A SPACE.

A space, or universe of discourse, U can be shown schematically as a rectangle. A statement in U can be suggested by an area containing exactly those objects x in U that make the statement have truth value True.

Since the rules of statement logic are to apply without regard to the meaning of the statements, we have to make two requirements of a sentence if it is to be a statement in the space U. A sentence is a **statement in U,** if:

1. it concerns the objects in U, and
2. it can take on exactly one of two **truth values, True** or **False,** for each object in U.

It is not the truth of a sentence that makes it a statement; we may know it to be False or we may not know its truth value, but we must know that its truth value is "decidable" and in just one way for each object in U.

Let $P(x)$ be a statement in a space U, such as "x is over 30" in the space "people now in this building" (Fig. 3–2). The objects x in U for which $P(x)$ takes on the value True can be pictured schematically within the box U by an area, called P. (We have drawn it to suggest a capital letter P.) Then the area inside the P-shaped curve is the "truth set" for the statement $P(x)$ made up of all the x's in U for which $P(x)$ is True. Since $P(x)$ is either True or False for each object in all of U, each object in U is within P or outside P, with no borderline cases.

The restriction to True or False statements, necessary here to formulate rules for deduction, is familiar from guessing games. Guessers are

FIGURE 3–2. Truth value of a statement in a space.

A statement in a space U has truth value True or False for each element of U. The area P represents those elements for which the statement $P(x)$ has value True. There are no borderline cases, because the statement can have only one truth value for each element.

restricted to questions that have just one correct answer, "yes" or "no." Questions such as "Is it big?" or "What color is it?" have to be rephrased as "Is it between the size of an ordinary dried pea and a football?" or "Is it in the red part of the spectrum?"

EXAMPLE   Let the space U be various breeds of dogs. The following sentences are statements in U, because each concerns breeds of dogs, and each has one truth value, either True or False, for each x in U:

> x was developed in Germany.
>
> x is a recognized breed of the American Kennel Club.
>
> x is not a terrier.

The sentence

> x is higher than middle C

is not defined over U. Although it makes perfectly good sense in the space of musical notes, it is nonsense in U, and so is not a statement in U.
    The sentence

> x is a good breed

is not definite enough to be decidable for every x in U, and so is not a statement in U.

Exercise 3–1. Discuss the idea of a statement in a space U in connection with "relevance."

Exercise 3–2. Let the space U be made up of six people:

$U = \{$Abraham Lincoln, Franklin D. Roosevelt, John Woolman, Louisa May Alcott, Ralph Waldo Emerson, Johannes Brahms$\}$.

Which of the following three sentences are statements in U? Which fails to be decidable?

> x was a president of the United States
>
> x influenced legislation
>
> x was a man

Exercise 3–3. In Figure 3–3 the six objects constitute a space U. Let P refer to the sentence

> $P(x)$: x is made of wood

Supposing there is glass in the window, is P a statement in U? Now exclude the window and show that P is a statement in the more restricted space V of the other five objects. Find each truth value.

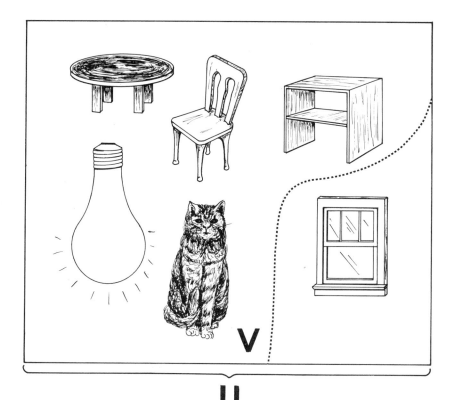

FIGURE 3–3. A space U of six objects, and a more restricted space V of five objects.

A sentence may fail to be a statement in a space U but become a statement in a more restricted space V where it has a truth value for every element.

Exercise 3–4. Let the space U consist of the objects shown in Figure 3–4. Tell whether each of the following sentences is a statement in U, and why.

(a) x is alive.
(b) x might be found in a kitchen.
(c) x is sharp.
(d) x is pretty.
(e) x is a banana.
(f) x is made of metal.
(g) x has holes in it.
(h) $x + 2$ is 15 or more.

FIGURE 3–4. THE SPACE U REFERRED TO IN EXERCISE 3–4.

Exercise 3–5. Rephrase each of the following questions as a statement in the space of objects in a room:

What color is it?
Where is it located in the room?
Is it a person or touching a person?

STATEMENT CONNECTIVES

There are several "connectives" that make new statements from old ones, just as the connective English word "and" makes the compound sentence

"Alaska is the westernmost state *and* Alaska is the easternmost state."

from the two statements

"Alaska is the westernmost state"

and "Alaska is the easternmost state."

In fact, the connective introduced in Figure 3–5 is just the formal "and" for statements in a logic system.

Because we are studying statements in general, without regard to their individual meanings, we rely solely on their truth values. That is why the definition of the compound statement $P(x) \wedge Q(x)$ is given entirely by the truth table, which shows the truth value for each combination of values for $P(x)$ and $Q(x)$ separately.

Symbols	Truth Table			English Expressions	
	$P(x)$	$Q(x)$	$P(x) \wedge Q(x)$	P and Q	
	T	T	T	P, but Q	
	T	F	F	P, moreover	
$P(x) \wedge Q(x)$	F	T	F	Q	
	F	F	F	P, also Q	
				P,Q, too	Diagram with x's shaded that make $P(x) \wedge Q(x)$ True

FIGURE 3–5. THE CONNECTIVE $\wedge$ FOR STATEMENTS $P(x)$, $Q(x)$ IN A SPACE U.

For any x that makes both $P(x)$ and $Q(x)$ True, the compound $P(x) \wedge Q(x)$ is True. All other x's make $P(x) \wedge Q(x)$ False.

Let $P(x)$ be "x has brothers." Let $Q(x)$ be "x has sisters." Then **EXAMPLE** $P(x) \wedge Q(x)$ stands for "x has both brothers and sisters."

Exercise 3–6. Adopt a space U and write out statements $P(x)$ and $Q(x)$ so that $P(x) \wedge Q(x)$ will read "x is a gentleman and a scholar." Draw a diagram and describe the people in each of the four areas. Are women included? Where?

Exercise 3–7. Let $P(x)$ be "x has brains" and $Q(x)$ be "x has beauty." State $P(x) \wedge Q(x)$, and make a truth table showing the truth value of the compound statement when x is brainy and beautiful, when x is brainy but not beautiful, and so on.

Exercise 3–8. If $R(x)$ is "x is driving" and $S(x)$ is "x is drinking," translate $R(x) \wedge S(x)$.

Exercise 3–9. Translate $Q(x) \wedge T(x)$ for the statements $Q(x)$: "x is frumious" and $T(x)$: "x is a bandersnatch," regardless of the meanings of the statements, if any.

Exercise 3–10. Write out statements $P(x)$ and $Q(x)$ and express with a logical statement connective: "x was small, but he was strong." (It is not unusual to lose shades of meaning in our very formal statement logic.)

Exercise 3–11. In Figure 3–6 let $M(x)$ be "x is a point in Main Street," and let $P(x)$ be "x is a point in Park Street." Use the connective $\wedge$ to describe those points that are in the intersection at Main and Park.

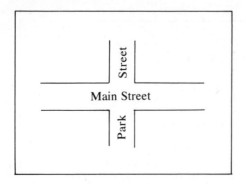

FIGURE 3–6. Illustration for Exercise 3–11.

The connective "∨" introduced in Figure 3–7 is the formal "or" used in logic.

Symbols	Truth Table			English Expressions	
	$P(x)$	$Q(x)$	$P(x) \vee Q(x)$	*P* or *Q*	
	T	T	T	either *P* or *Q*	
$P(x) \vee Q(x)$	T	F	T	one of *P*, *Q*	
	F	T	T	either *P* or *Q*,	
	F	F	F	or both	
				"inclusive" *P* or *Q*	

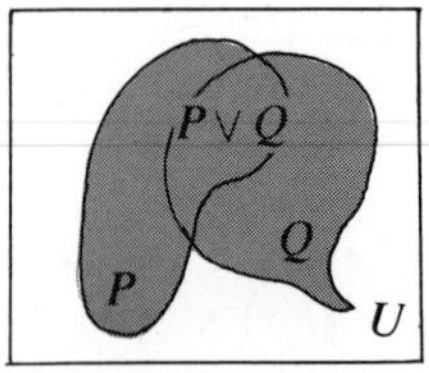

Diagram with *x*'s shaded that make $P(x) \vee Q(x)$ True

FIGURE 3–7. The connective ∨ for statements $P(x)$, $Q(x)$ in a space U.

The form of "or" used in the logical "∨" is inclusive; that is, $P(x) \vee Q(x)$ has truth value True when one of $P(x)$, $Q(x)$ is True and also when both are True. It is False only when both $P(x)$ and $Q(x)$ are False.

EXAMPLE   Let $P(x)$ be "x needs a jug of wine," and let $Q(x)$ be "x needs a loaf of bread." Then $P(x) \vee Q(x)$ represents "x needs wine or bread or both."

Exercise 3–12. Write the statement $P(x) \vee Q(x)$ for the simple statements of Exercise 3–7. Show $P \vee Q$ on a diagram.

Exercise 3–13. A restaurant menu says, "Dinner includes soup or salad." Show on a diagram what is probably meant and contrast this with the meaning in logic of "Dinner includes soup" ∨ "Dinner includes salad."

Exercise 3–14. Let $P(x)$ be "x has blond hair" and $Q(x)$ be "x has brown eyes." Use the connective "$\vee$" and translate the resulting compound sentence.

Exercise 3–15. You are sorting student file cards according to sex and undergraduate or graduate status. Adopt symbolism and show the four combinations of truth values for the two basic statements.

The negation connective introduced in Figure 3–8 does not connect two statements, but provides the opposite of a statement $P(x)$. This opposite has truth value False in U wherever $P(x)$ is True and truth value True wherever $P(x)$ is False. Since in diagrams we have adopted the convention of letting the area enclosed in P stand for the objects of U that make $P(x)$ True, the outside portion stands for the objects that make $\sim P(x)$ True.

Symbols	Truth Table		English Expressions
	$P(x)$	$\sim P(x)$	not P
	T	F	deny that P
$\sim P(x)$	F	T	negation of P
			it is False that P
			opposite of P

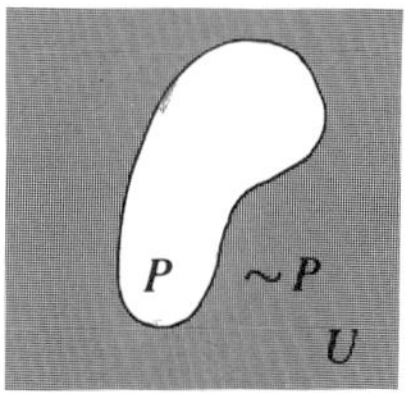

Diagram with x's shaded that make $\sim P(x)$ True

FIGURE 3–8. THE NEGATION $\sim$ FOR A STATEMENT $P(x)$ IN A SPACE U.

Let $P(x)$ be the statement in the space of wigs: "x costs more than **EXAMPLE** \$15." Then $\sim P(x)$ is the statement "x does not cost more than \$15." A wig costing \$21 falls in area P in the diagram. A wig costing \$14.95 falls outside P. A wig costing exactly \$15 falls outside P. No objects in U fall on the borderline, since each makes $P(x)$ True (and $\sim P(x)$ False) or False (and $\sim P(x)$ True).

Exercise 3–16. Write out the statement $\sim P(x)$ if $P(x)$ is the statement in the space of trees: "x is an evergreen." Draw a diagram and make a truth table showing P and $\sim P$.

Exercise 3–17. You have the negative and a print of a black-and-white photograph. Let $B(x)$ for small areas x on the photograph be "x is black." Describe the negative and the print, using $B(x)$ and $\sim B(x)$.

Some shades of meaning of the English word "not" cannot be translated completely by "∼." For instance, "x does *not* like cucumbers" is not precisely the negation of "x likes cucumbers," because it leaves out x's who are indifferent to cucumbers. Ordinary English expression is not restricted to True-or-False statements, but allows for three or more valued logics. To translate ∼P(x) to English when P(x) is "x likes cucumbers," say, "It is false that x likes cucumbers."

Exercise 3–18. Write the statement ∼P(x), if P(x) is the statement in the space of people "x likes maple walnut ice cream." Make a diagram and a truth table showing P and ∼P.

Before introducing other connectives, we present some examples and exercises involving combinations of ∧, ∨, and ∼.

EXAMPLE   Let U be the space of people with various personality traits. Let P(x) be the statement "x is clever" and let Q(x) be "x is trustworthy." Then in Figure 3–9, area P encloses all the people x for whom P(x) has value True, that is, all people who are clever. Area Q encloses all people x for whom Q(x) is True, that is, the x's that make Q a True statement. Then the area common to both areas represents P ∧ Q, and is made up of those people who are both clever and trustworthy. The shaded area represents (∼P) ∨ Q, the people who are not clever or are trustworthy or both. As the accompanying truth table shows, (∼P) ∨ Q is True for three of the four combinations of truth values for P and for Q.

The truth table for (∼P) ∨ Q in Figure 3–9 is completed in two stages.

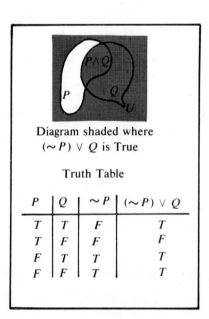

Diagram shaded where
(∼P) ∨ Q is True

Truth Table

P	Q	∼P	(∼P) ∨ Q
T	T	F	T
T	F	F	F
F	T	T	T
F	F	T	T

FIGURE 3–9. TRUTH TABLE FOR (∼P) ∨ Q.

First, fill in the column for "~P." Then using only the "~P" column and the Q column, connect them with "v."

Exercise 3–19. Make a diagram and a truth table for the compound statement ~(P v Q), and use them to show that the statement ~(P v Q) is different from the statement (~P) v Q shown in Figure 3–9. Use the example of clever and trustworthy people in comparing English translations.

Exercise 3–20. Let $P(x)$ be "x is a bird" in the space of living creatures, and let $Q(x)$ be "x can swim." Draw a diagram and construct a truth table showing (~P) v (~Q). Consider the object "penguin" in U. What truth values does this particular x give to $P(x)$ and to $Q(x)$? Then where do penguins lie in the diagram of the statements in U.

Exercise 3–21. In Figure 3–10 find which cars x make each of the following statements True.

$P(x)$: x is a convertible.
$Q(x)$: x is a two-seater. (Rumble seats don't count.)
$R(x)$: x has its top down in the picture.
$S(x)$: x is a sedan (has both front and back seats).
~P(x), P v Q, P ∧ S, (~P) ∧ (~S), P ∧ (~R).

1932 Ford

1930 Cadillac — La Salle

1933 DeSoto

1926 Buick

1928 Packard

FIGURE 3–10. SPACE OF CARS FOR EXERCISE 3–21.

Exercise 3–22. Let $P(x)$ be "x is a student" and let $Q(x)$ be "x is a radical in politics." Adopt a space U and make a diagram and a truth table showing $(\sim P) \vee (\sim Q)$. What individuals x are *not* in $(\sim P) \vee (\sim Q)$?

Exercise 3–23. Make a diagram and a truth table showing $P \vee (\sim Q)$. Let $P(x)$ be "x costs a round number of dollars (no change required)" and let $Q(x)$ be "x costs a round number of ten-dollar bills." Are there any x's (in the space of priced objects) that are *not* in $P \vee (\sim Q)$; that is, x's that make $P(x) \vee [\sim Q(x)]$ False for these interpretations of $P(x)$ and $Q(x)$?

Exercise 3–24. Compare $\sim(P \wedge (\sim Q))$ with the statement $(\sim P) \vee Q$, by making diagrams and truth tables for both of them. Compare "it is false that a student can do his homework and still fail the course" and "either a student does not do his homework or he passes the course."

Figure 3–11 introduces the conditional, $P(x) \rightarrow Q(x)$, often expressed in English as "If $P(x)$, then $Q(x)$." As the truth table shows, however, the use of the conditional in logic differs from some uses of "If . . . , then . . . ," in that it covers the case of False "If . . ." clauses. This is because our definitions must cover every contingency, since we do not know the content of the statements we are working with, and so do not know whether they are True or False but only that they are one or the other unambiguously. In case $P(x)$ is False, $Q(x)$ need not be True. It can have either truth value. Only if $P(x)$ has value True is there a requirement on $Q(x)$ when $P(x) \rightarrow Q(x)$ is True. The logical conditional $P(x) \rightarrow Q(x)$ carries with it no connotation of causation. A conditional in ordinary English often does carry a suggestion of causation: "If you fall off that cliff, you will break your neck," for example. In the formal logic system the conditional is True as long as $Q(x)$ is True whenever $P(x)$ is True, even if it is mere coincidence.

Symbols	Truth Table			English Expressions	
	$P(x)$	$Q(x)$	$P(x) \rightarrow Q(x)$	if P, then Q	
$P(x) \rightarrow Q(x)$	T	T	T	P implies Q	
	T	F	F	Q if P	
	F	T	T	Q provided that P	
	F	F	T	P is a sufficient condition for Q	
				P only if Q	

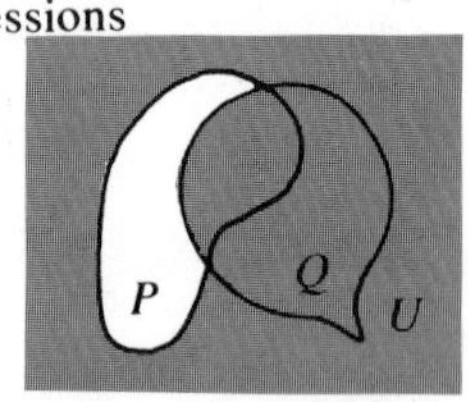

Diagram with x's shaded that make $P(x) \rightarrow Q(x)$ True

FIGURE 3–11. THE CONNECTIVE $\rightarrow$.

EXAMPLE In "If you need me, you will find me in the library," the $Q(x)$ clause, "you will find me in the library," may be True whether "you need me" or not. Then the conditional is True.

Let $P(x) \rightarrow Q(x)$ be "If you study engineering, then you need lots of **EXAMPLE** math." The three True combinations are: studying engineering and needing math, not studying engineering and still needing math, and not studying engineering and not needing math. The only excluded possibility if the conditional is True is engineering-and-no-math.

Exercise 3–25. Let $P(x)$ be "x is smart" and let $Q(x)$ be "x is lucky." Translate the conditional $P(x) \rightarrow Q(x)$. Make a diagram and a truth table.

Exercise 3–26. Write as a conditional: "If you doze in class, you need more sleep at night." Write as a conditional using the same notation: "If you need more sleep at night, you will doze in class."

Exercise 3–27. Let $P(x)$ be "x is a wet day" and let $Q(x)$ be "x is a windy day." Prepare a diagram and a truth table showing the truth value of the conditional $P(x) \rightarrow Q(x)$ for wet windy days, dry windy days, wet calm days, and dry calm days. Is the truth value of the conditional found from studying weather facts?

Exercise 3–28. Let $P(x)$ be "x likes people," $Q(x)$ be "x likes dogs," and $R(x)$ be "x has lots of friends." Translate $[P(x) \wedge Q(x)] \rightarrow R(x)$.

Exercise 3–29. Let $P(x)$ be "x fights" and $Q(x)$ be "x runs away" and $R(x)$ be "x lives to fight another day." Express in symbolism "He who fights and runs away lives to fight another day."

Exercise 3–30. Let $P(x)$ be "x understands," and let $Q(x)$ be "x asks questions." Express in symbolism "If x does not understand, then either he asks questions, or else he will still not understand."

Exercise 3–31. In Figure 3–4, let $P(x)$ be "x is entirely edible," and let $Q(x)$ be "x is entirely fruit." Determine which x's in the picture make the conditional $P(x) \rightarrow Q(x)$ True.

Exercise 3–32. In Figure 3–10, let $P(x)$ be "x is a convertible," and let $R(x)$ be "x has its top down in the picture." Determine which cars in the picture make the conditional $P(x) \rightarrow R(x)$ True.

Exercise 3–33. Write in symbols "If it isn't a bird or a plane, it is Superman."

Exercise 3–34. Let the space U be the whole numbers from 1 to 10. Let $P(x)$ be "x is greater than 3," and let $Q(x)$ be "x is greater than 5." Find which of the ten x's make the conditional $P(x) \rightarrow Q(x)$ True.

Figure 3–12 introduces our last sentence connective, the biconditional, which in English often takes the form "$P(x)$ if and only if $Q(x)$."

Symbols	Truth Table			English Expressions
	$P(x)$	$Q(x)$	$P(x)\leftrightarrow Q(x)$	P if and only if Q
$P(x)\leftrightarrow Q(x)$	T	T	T	P is a necessary
	T	F	F	and sufficient
	F	T	F	condition for Q
	F	F	T	if P then Q, and

if Q then P Diagram with x's shaded that make $P(x)\leftrightarrow Q(x)$ True

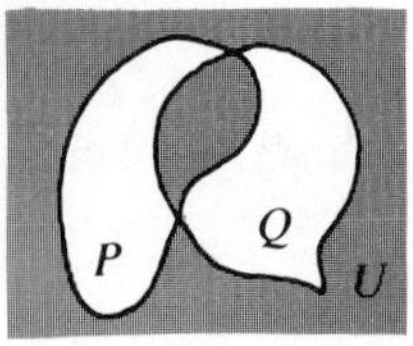

FIGURE 3–12. THE CONNECTIVE $\leftrightarrow$.

EXAMPLE   As in Exercise 3–34, let the space U be the whole numbers from 1 to 10. Let $P(x)$ be "x is greater than 3," and let $Q(x)$ be "x is greater than 5." According to Figure 3–12, the biconditional $P(x) \leftrightarrow Q(x)$ is True when $P(x)$ and $Q(x)$ have the same truth value. This can be determined directly for the ten x's in U:

x	x is greater than 3	x is greater than 5	$P(x) \leftrightarrow Q(x)$
1	F	F	T
2	F	F	T
3	F	F	T
4	T	F	F
5	T	F	F
6	T	T	T
7	T	T	T
8	T	T	T
9	T	T	T
10	T	T	T

The biconditional is True for 1, 2, 3, 6, 7, 8, 9, and 10, but False for 4 and 5.

Exercise 3–35.   In Figure 3–10, find which cars make the biconditional True: $Q(x) \leftrightarrow R(x)$, where $Q(x)$ is "x is a two-seater," and $R(x)$ is "x has its top down in the picture."

Exercise 3–36.   Let the space U be the following male-female pairs:

billy-nanny (goats)
jack-jenny (asses)
drake-duck
gander-goose
rooster-mare
lion-bitch

For each pair x, let $P(x)$ be "x has as its first listing the male name for animal A," and let $Q(x)$ be "x has as its second listing the female name for animal A." Find which pairs x make the biconditional True.

Exercise 3–37. At the end of a dinner, all the men have finished and all the women are still eating. Adopt symbolism and write this information as a biconditional.

Exercise 3–38. All animals that have hair, suckle their young, and bear live young (call them x's for which $H(x)$ is True) are mammals (say, make $M(x)$ True). Except for the monotremes, all mammals make $H(x)$ True. Then which animals make the biconditional $H(x) \leftrightarrow M(x)$ False? (The monotremes, the duck-billed platypus and the spiny anteater, lay eggs.)

Exercise 3–39. In Figure 3–13 find where the conditional $P(x) \rightarrow S(x)$ is True and where it is False.

FIGURE 3–13. Statements about a tabletop arrangement.

In this tabletop arrangement, the space U is the space of models of wheeled transport. Three of the models shown are not in U, since they do not have wheels. The following names for statements in U appear in exercises:

$P(x)$: "x is a pleasure automobile."
$Q(x)$: "x is a commercial highway vehicle."
$R(x)$: "x runs on tracks."
$S(x)$: "x is an antique car."
$T(x)$: "x is driven only for transportation to work."

Exercise 3–40. Find where the conditional $S(x) \rightarrow P(x)$ is True and where it is False in Figure 3–13.

Exercise 3–41. In Figure 3–13 show in a truth table that the biconditional

$$[R(x) \lor T(x)] \leftrightarrow [(\sim P(x)) \land (\sim Q(x))]$$

has truth value True for every vehicle in U.

Exercise 3–42. In Figure 3–13 show that the conditional $(P(x) \land Q(x)) \rightarrow P(x)$ has truth value True for every combination of truth values for P and Q.

Exercise 3–43. In Figure 3–13 show that the conditional $R(x) \to$ $[R(x) \vee S(x)]$ has truth value True for every combination of truth values for R and S.

Exercise 3–44. In Figure 3–13 show that the biconditional

$$\sim(P(x) \vee Q(x)) \leftrightarrow [(\sim P(x)) \wedge (\sim Q(x))]$$

has truth value True for every combination of truth values for P and Q.

PATTERNS OF DEDUCTION

Most of our examples have concerned statements with truth value True for some objects in the space U, and with truth value False for others. However, a few statements, because of their form alone, have only True values with no False values. These statements are called tautologies.

A **tautology** in a space U is a statement that has truth value True for all combinations of values for the basic statements involved.

EXAMPLE   In the discussion on page 65 we mentioned that the statement B-and-C, which we now write in formal language "$B(x) \wedge C(x)$" or, more conveniently, simply "$B \wedge C$," seems to have the statement B folded within itself. Without knowing what words B and C stand for, or what truth values they have for various x's in U, we feel we can "deduce" the conclusion B from the hypothesis $B \wedge C$ by the form alone. We can show in a truth table like that of Figure 3–14 that the conditional $[B \wedge C] \to B$ is a tautology; that is, it has constant truth value True. To show this fact in symbols, we replace the conditional sign $\to$ by the sign $\Rightarrow$, which we read "**tautologically implies**" or "**formally implies.**"

Exercise 3–45. In the truth table of Figure 3–14, why is the conditional $[B \wedge C] \to B$ True whenever the "if" part, $B \wedge C$ is False? (See Figure 3–11.)

B	C	B ∧ C	(B ∧ C) → B	
T	T	T	T	
T	F	F	T	$(B \wedge C) \Rightarrow B$
F	T	F	T	
F	F	F	T	

FIGURE 3–14. THE TAUTOLOGY $(B \wedge C) \Rightarrow B$.

The conditional $(B \wedge C) \to B$ is a tautology, because all its truth values are True for all combinations of truth values for the basic statements B and C.

Exercise 3–46. Add a column to the truth table of Figure 3–14 to show that $(B \wedge C) \Rightarrow C$ is also a tautology.

Exercise 3–47. Compare the conditional in Exercise 3–42 with the tautology of Figure 3–14. Show that the conditional in Exercise 3–39 is not a tautology.

Since a tautology is *formally* valid, that is, True by its form alone without reference to the content in a specific example, it supplies a "pattern of deduction" that can be applied to any reasoning that is limited to statements in a space U.

If a recipe requires eggs and sugar, then it certainly requires eggs. **EXAMPLE** To show that this conditional has the tautological form of Figure 3–14, assign statement notations:

$$E(x): \text{``}x \text{ requires eggs''}$$

$$S(x): \text{``}x \text{ requires sugar''}$$

in the space of recipes. Then the "if" clause has the form $E(x) \wedge S(x)$, and the whole tautological conditional takes on the form

$$[E(x) \wedge S(x)] \Rightarrow E(x).$$

He is a scoundrel, but a likeable one. Translating the word "but" to **EXAMPLE** its closest formal symbol "$\wedge$," and choosing statement notation

$$S(x): \text{``}x \text{ is a scoundrel''}$$

$$L(x): \text{``}x \text{ is likeable''}$$

we can draw the pessimistic conclusion from the form of the given statement alone that "he is a scoundrel," or, following Exercise 3–46, we can draw the optimistic conclusion that "he is likeable."

$$[S(x) \wedge L(x)] \Rightarrow S(x) \quad \text{and also} \quad [S(x) \wedge L(x)] \Rightarrow L(x).$$

Exercise 3–48. Express in formal terms and draw a formal conclusion from the statement "That required book is long and dull" in the space of required books.

Exercise 3–49. Make an eight-row truth table showing that this statement involving three basic statements is a tautology: $(P \wedge Q \wedge R) \Rightarrow R$.

Exercise 3–50. Draw two formal conclusions from the statement: "This person is a young hoodlum."

In Exercise 3–43 the conditional $R(x) \rightarrow [R(x) \vee S(x)]$ was shown to have truth value True throughout a space of car models in which $R(x)$ and

$S(x)$ represented specific statements. That conclusion was independent of the car model setting, for the conditional is a tautology, as shown in Figure 3–15.

R	S	R v S	R → (R v S)
T	T	T	T
T	F	T	T
F	T	T	T
F	F	F	T

$R \Rightarrow (R \vee S)$

FIGURE 3–15. THE TAUTOLOGY $R \Rightarrow (R \vee S)$.

The conditional $R \rightarrow (R \vee S)$ is True for all combinations of the basic statements R and S, which proves that the conditional is a tautology.

Exercise 3–51. When the "if" clause R is False in Figure 3–15, sometimes the "then" clause $(R \vee S)$ is False and sometimes it is True. Point out where the truth table in Figure 3–15 shows this and explain why the conditional in the last column is True in either case.

As a tautology, then, the statement pictured in Figure 3–15 constitutes another valid pattern of reasoning, and can be applied whenever the context is formal enough to provide sentences that are statements in a space.

EXAMPLE   "She is a winner" implies "Either she is a winner or I'll eat my hat," and the implication is completely formal, depending for its validity on the pattern alone without reference to content or my appetite for hats.

Exercise 3–52. Translate the tautology $R(x) \rightarrow [R(x) \vee S(x)]$ for statements

$R(x)$: "You are sentenced to 30 days in jail."

$S(x)$: "You are sentenced to pay a fine of $100."

What happens if we exchange the "if" clause with the "then" clause in the conditional of Figure 3–15? The statement $R(x) \vee S(x) \rightarrow R(x)$ is not a valid pattern of reasoning, as can be shown by a space in which it has False values for some combinations of basic statement values. For instance, let the space U be four books on my shelf: *Birds of the World*, *The International Atlas*, *Webster's Unabridged Dictionary*, and *Roget's Thesaurus*. Let $R(x)$ be "x is a mathematics book," and let $S(x)$ be "x is a reference book." Then $R(x) \vee S(x)$ is True for each x in U, but $R(x)$ is False, making the conditional False. To be a tautology, the conditional would have to have constant truth value True.

The conditional $Q \rightarrow P$ is called the **converse** of the conditional

$P \rightarrow Q$. We have just shown that the converse of a tautological conditional is not necessarily a tautology.

Exercise 3–53. Show by inventing an example that the converse of the tautology of Figure 3–14 is not a tautology.

One common mistake in reasoning is to replace a conditional unintentionally by its converse. In fact, one technique of propaganda is to create confusion between a readily granted statement like "If you have a new car you will be happy" and its converse "If you are to be happy, you will have to have a new car."

Exercise 3–54. Show by a truth table that the statement

$$P \vee \sim P \text{ (The \textbf{Law of the Excluded Middle})}$$

is a tautology. Explain how the tautology comes directly from part of the definition of a statement.

Exercise 3–55. Show by a truth table that

$$\sim[P \wedge (\sim P)] \text{ (The \textbf{Law of Contradiction})}$$

is a tautology, and explain how it comes directly from part of the definition of a statement. (Also see Exercise 3–54.)

Equivalence

If two statements are connected by a biconditional that is a tautology, written

$$P \Leftrightarrow Q,$$

then we call P and Q **equivalent** statements and use them interchangeably, substituting one for the other freely.

For example, we can show that $\sim(\sim P)$ is equivalent to P, because the biconditional $\sim(\sim P) \Leftrightarrow P$ is a tautology. This is established by the truth table shown in Figure 3–16.

P	$\sim P$	$\sim(\sim P)$	$\sim(\sim P) \leftrightarrow P$
T	F	T	T
F	T	F	T

FIGURE 3–16. $\sim(\sim P)$ IS EQUIVALENT TO P.

The statements $\sim(\sim P)$ and P are equivalent because their biconditional $\sim(\sim P) \leftrightarrow P$ is a tautology.

Exercise 3–56. Translate the equivalence shown in Figure 3–16 if

$\sim(\sim P)$ stands for "He was *not impolite*." Is the equivalence exact in everyday English usage of the words "impolite" and "not polite"? Are "polite" and "impolite" opposites in the sense that the statement "he was impolite" always has the opposite truth value from the statement "he was polite"?

We can express some of the five connectives $\wedge$, $\vee$, $\sim$, $\rightarrow$, and $\leftrightarrow$ by means of equivalent statements using just two of them. For example, Figure 3–17 shows that $P \wedge Q$ is equivalent to $\sim[(\sim P) \vee (\sim Q)]$, so that we could avoid the symbol $\wedge$ if we were content with the somewhat clumsier equivalent using $\vee$ and $\sim$.

P	Q	$\sim P$	$\sim Q$	$(\sim P) \vee (\sim Q)$	$\sim[(\sim P) \vee (\sim Q)]$	$P \wedge Q$	$(P \wedge Q) \leftrightarrow \sim[(\sim P) \vee (\sim Q)]$
T	T	F	F	F	T	T	T
T	F	F	T	T	F	F	T
F	T	T	F	T	F	F	T
F	F	T	T	T	F	F	T

FIGURE 3–17. An equivalent statement for $P \wedge Q$, using $\vee$ and $\sim$.

It is possible to represent the statement $P \wedge Q$ by an equivalent statement that uses $\vee$ and $\sim$ as the only connectives.

As shown in Figure 3–18, the conditional $P \rightarrow Q$ can be written in the equivalent form $(\sim P) \vee Q$.

P	Q	$\sim P$	$(\sim P) \vee Q$	$P \rightarrow Q$	$(P \rightarrow Q) \leftrightarrow [(\sim P) \vee Q]$
T	T	F	T	T	T
T	F	F	F	F	T
F	T	T	T	T	T
F	F	T	T	T	T

FIGURE 3–18. An equivalent statement for $P \rightarrow Q$, using $\vee$ and $\sim$.

A conditional statement can be expressed using $\vee$ and $\sim$ as the only connectives.

As shown in Figure 3–19, a biconditional is equivalent to a conditional and its converse connected by $\wedge$.

Since Figure 3–19 shows that a biconditional can be written in terms of conditionals and $\wedge$, and since Figure 3–18 shows that conditionals can be written in an equivalent form using only $\vee$ and $\sim$, and since Figure 3–17 shows that $\wedge$ can be replaced by $\vee$ and $\sim$, it is possible to replace a biconditional by an equivalent statement involving only $\vee$ and $\sim$.

P	Q	P → Q	Q → P	$(P \rightarrow Q) \wedge$ $(Q \rightarrow P)$	P ↔ Q	$(P \leftrightarrow Q) \leftrightarrow [(P \rightarrow Q) \wedge$ $(Q \rightarrow P)]$
T	T	T	T	T	T	T
T	F	F	T	F	F	T
F	T	T	F	F	F	T
F	F	T	T	T	T	T

FIGURE 3–19. BICONDITIONAL EQUIVALENT TO CONDITIONAL AND ITS CONVERSE CONNECTED BY ∧.

A biconditional can be represented by an equivalent statement using → and ∧.

A **definition** in mathematics is always a tautological biconditional, an "if and only if" statement that is True in the whole space. Take as an example the following definition:

Definition: A rectangle is a square if and only if its length and width measurements are equal.

We could abbreviate this "$S(x) \Leftrightarrow M(x)$: In the space of rectangles, x is a square if and only if x has equal measurements of length and width." Notice that oblong rectangles make both the "if" part and the "then" part False, and the biconditional True.

As another example, consider the definition we have been using for a logical statement:

Definition: A sentence is a statement in a space if (1) it is about objects in the space, and (2) it has exactly one truth value for each object in the space.

Although this definition has only the word "if" in it, the title "Definition" carries also an "only if" restriction. A sentence is a statement *if* the two requirements are met, and it is a statement *only if* the two requirements are met. Some authors write this tautological "if and only if" as **"iff."** If the tautology $P \Leftrightarrow Q$ holds, then Q may be said to characterize P completely and so define it. Even this definition of a definition is a tautological biconditional! That is, it is True throughout formal mathematics (a tautology) that a statement Q can serve as definition of P *if and only if* $P \Leftrightarrow Q$ is a tautology. Also, if Q defines P then P satisfies all the requirements necessary to make it a definition of Q.

Exercise 3–57. A biologist may refer to bears by their family name Ursidae, because the name "bear" sometimes refers to honey bears, koala bears, or wooly bears. Compare scientific names with common names. Is the name "bear" too inclusive or not inclusive enough; that is, does the "if" or the "only if" fail between "x is a bear" and "x is of the family Ursidae"?

Exercise 3–58. If a book is an autobiographical novel, does it have to be a novel? If a book is autobiographical, does it have to be an auto-biographical novel?

Exercise 3–59. Use a truth table to establish the tautology $\sim(\sim P) \Leftrightarrow P$. Check your work by reference to Figure 3–16.

Exercise 3–60. Use a truth table to establish the tautology $(P \wedge Q) \Rightarrow P$.

Exercise 3–61. Show that $\sim(P \vee Q)$ is equivalent to $(\sim P) \wedge (\sim Q)$. (**De Morgan* Law**)

Exercise 3–62. Show that $\sim(P \wedge Q)$ is equivalent to $(\sim P) \vee (\sim Q)$. (**De Morgan Law**)

Exercise 3–63. Establish the tautology $(P \rightarrow Q) \Leftrightarrow [(\sim Q) \rightarrow (\sim P)]$.

The conditional $(\sim Q) \rightarrow (\sim P)$ is called the **contrapositive** of the conditional $P \rightarrow Q$. As shown in Exercise 3–63, the contrapositive is equivalent to the **direct** conditional $P \rightarrow Q$ and so can be substituted for it. This is especially helpful if the direct form would require checking many cases.

EXAMPLE   Suppose we want to establish that "If he is still here, he is interested in the subject." Instead of checking everyone still here, we can use the contrapositive idea that anyone not interested in the subject would have left long ago. If he is not interested, then he is not still here.

Exercise 3–64. Establish the tautology $(Q \rightarrow P) \Leftrightarrow [(\sim P) \rightarrow (\sim Q)]$. You can use the previous exercise to establish this tautology by arguing that Exercise 3–63 holds for any statements P and Q, in particular for Q and P.

Recall that $Q \rightarrow P$ is the converse of the direct conditional $P \rightarrow Q$. The conditional $(\sim P) \rightarrow (\sim Q)$ is called the **inverse** of $P \rightarrow Q$.

direct $P \rightarrow Q \Leftrightarrow$ contrapositive $(\sim Q) \rightarrow (\sim P)$
converse $Q \rightarrow P \Leftrightarrow$ inverse $(\sim P) \rightarrow (\sim Q)$

FIGURE 3–20. FOUR CONDITIONALS.

The four conditionals that can be formed with statements P and Q are related in two pairs of biconditionals.

---

* De Morgan (1806–1871) was a pioneer in the development of a formal logic, preparing the way for his contemporary George Boole (1815–1864), and his "Laws of Thought."

Translate the direct conditional **EXAMPLE**

$P \rightarrow Q$: "If I'd listened to what Maw said, I'd be sleeping in a featherbed."

This is equivalent to the contrapositive

$(\sim Q) \rightarrow (\sim P)$: "If I'm not sleeping in a featherbed, then I didn't listen to what Maw said." ·

The converse, which is not equivalent to $P \rightarrow Q$, is

$Q \rightarrow P$: "If I'm sleeping in a featherbed, then I listened to what Maw said."

The converse is equivalent to the inverse,

$(\sim P) \rightarrow (\sim Q)$: "If I didn't listen to what Maw said, then I'm not sleeping in a featherbed."

Exercise 3–65. Let $P \rightarrow Q$ be "If I am hungry, then I eat." Write the inverse, the converse, and the contrapositive, and show which are equivalent.

Exercise 3–66. Let $(\sim P) \rightarrow (\sim Q)$ be "If you won't marry me, then I cannot be happy." Take this to be the inverse conditional and write the direct conditional, the contrapositive, and the converse.

Exercise 3–67. Establish by a truth table the tautology

$$[P \wedge (P \rightarrow Q)] \Rightarrow Q. \quad \text{(The \textbf{Law of Detachment})}$$

In debate, the law of Detachment is often called by its Latin name *modus ponens*. It is this pattern of deduction we use to conclude Q from the conditional $P \rightarrow Q$ whenever we know P to be True. For instance, we might know the conditional "If an animal is in the cat family, then it is a mammal" to be True throughout U. Then whenever an animal is known to be in the cat family, such as cheetahs, ocelots, tigers, we can deduce by the Law of Detachment that it is a mammal.

Exercise 3–68. Use the Law of Detachment to draw a conclusion from "If there's red sky in the morning, then sailors take warning" and "There is a red sky this morning."

Exercise 3–69. Complete this argument to present an application of the Law of Detachment: A Boy Scout is a friend to animals, so Roy must be a friend to animals.

Exercise 3–70. Show how the Law of Detachment is used in law: For instance, suppose a law says that if a defendant is convicted of fraud, then he will be sentenced to one-to-three years in prison. A trial establishes which part of the tautology?

Exercise 3–71. If an animal is a quadruped, then it has four legs. A table has four legs, so a table is a quadruped. Show that this argument does not follow the pattern of the Law of Detachment.

Exercise 3–72. An old song says, "If I'm not near the girl I love, I love the girl I'm near." Suppose you *are* near the girl you love. Does the Law of Detachment apply?

Another important pattern of deduction is the **syllogism** shown in the next exercise.

Exercise 3–73. Establish the tautology $[(P \to Q) \wedge (Q \to R)] \Rightarrow (P \to R)$.

EXAMPLE   Take as the **hypothesis** of the syllogism the two **premises** "If I eat a salted peanut, I always eat a second one" and "If I eat two salted peanuts, I always eat a third one." This syllogism has the **conclusion**:

"If I eat one salted peanut, I eat three of them."

Like the other tautologies, the syllogism is True for all truth values of its basic statements, so it has formal validity, even if some of the premises are unusual.

EXAMPLE   Take as hypotheses the premises: "If a giraffe drinks ink, it insists on a drinking straw," and "If a giraffe insists on a drinking straw, it gets sick." In this case both premises would be hard to defend, while the conclusion appears comparatively reasonable: "If a giraffe drinks ink, it gets sick." The status of the premises and the conclusion, however, has nothing to do with the validity of the deduction, which is perfectly correct.

Exercise 3–74. Adopt notation and write this argument symbolically as a syllogism: If you practice, your work will improve, and if your work improves, you will enjoy it more. Therefore, if you practice, you will enjoy your work more.

Exercise 3–75. Take as premises of a syllogism: "If you hit me, I'll scream," and "If I scream, the neighbors will come to see what is wrong." What is the conclusion?

Exercise 3–76. You have a premise: "Very young girls have more finger dexterity than very young boys," and you want to conclude "Very young girls learn faster than very young boys." What other premise do you need?

Exercise 3-77. You state a premise: "The MacMahons are Scotch," and conclude: "Therefore, the MacMahons are thrifty." What premise is used but not stated?

Exercise 3-78. Given the premise "If you get the job, you start on Monday," what other premise do you need to conclude "If you make a good impression, you start on Monday?"

Exercise 3-79. Rely entirely on the structure of a formal syllogism to find the conclusion of this syllogism, without looking up unfamiliar terms in the premises: "If an equation is solvable by radicals, then there is a root tower of normal extensions for an extension field containing the solutions over the coefficient field. If there is a root tower of normal extensions for an extension field containing the solutions over the coefficient field, then the Galois group is solvable."

Exercise 3-80. Find the conclusion of this syllogism, without looking up unfamiliar terms: "A platyceromys is a jerboa; and a jerboa is of the family Dipodidae."

Exercise 3-81. Find the conclusion of this syllogism: "Crocidolite is a hornblende, and hornblende is a mineral."

Exercise 3-82. Suppose that the following nonsense constitutes premises of a syllogism, and find the conclusion: "If snoodge, then clorp, and if clorp, then smudgem."

Exercise 3-83. Show that this argument does not have the form of a syllogism: "MacSnort is a member of the Save the Garfish League. Three known Communists are members of the Save the Garfish League. Therefore, MacSnort is a Communist."

Exercise 3-84. Show that this argument does not have the form of a syllogism: "Miller is a mathematician, and Miller is rich; therefore, mathematicians are rich."

Exercise 3-85. Show that this ancient "paradox" has the form of a syllogism. Can you find a flaw in it? "A penny is better than nothing, and nothing is better than heaven; therefore, a penny is better than heaven."

Exercise 3-86. Use a truth table to prove that a syllogism with three premises is a tautology: $[(P \rightarrow Q) \wedge (Q \rightarrow R) \wedge (R \rightarrow S)] \Rightarrow (P \rightarrow S)$.

Exercise 3-87. Extend the previous exercise to cover four premises.

Exercise 3-88. Prove by a truth table the tautology

$$(H \rightarrow C) \Leftrightarrow \sim[H \wedge (\sim C)].$$

Indirect proofs are based on the tautology of Exercise 3–88, where H is a hypothesis (usually made up of several premises) and C is a conclusion. To prove that H logically implies C, that is $H \Rightarrow C$ is a tautology, we show in an indirect proof that $H \wedge (\sim C)$ logically implies a contradiction, that is, a statement having constant truth value False. For example, take as hypothesis the conditional statement:

"If Mary goes to town, then George and Larry must ride shotgun with her to protect her from bandits,"

and the statement

"George must stay home to practice his clarinet."

To show that this hypothesis implies the conclusion:

"Mary does not go to town,"

suppose the conclusion is False; that is, suppose Mary *does* go to town, even though the hypothesis holds (is tautologically True). From the first premise, since Mary goes to town, George and Larry ride with her (deduced by the Law of Detachment). Since George and Larry ride with her, George rides with her (deduced from the tautology $G \wedge L \Rightarrow G$). Then George rides to town and also stays home, which is a contradiction and has constant value False (from the Law of Contradiction). In symbols we could write the hypothesis $M \Rightarrow (G \wedge L)$ and $\sim G$, and the conclusion $\sim M$.

EXAMPLE   Take as hypothesis:
1. Henry climbs mountains only in the spring.
2. When Eva visits Manhattan, Henry climbs mountains.
3. It is not spring.

Prove the conclusion: Therefore, Eva is not in Manhattan.

Let P be "Henry climbs mountains,"
    Q be "It is spring,"
    R be "Eva visits Manhattan."

The hypothesis becomes

$P \Rightarrow Q$
$R \Rightarrow P$
$\sim Q$

and the conclusion becomes $\sim R$.

Then suppose $\sim\sim R = R$, that is, suppose R is tautologically True. All premises are assumed tautologically True. Then from premise 2, P is formally True by the Law of Detachment. Then from premise 1, Q, by the Law of Detachment. But $\sim Q$, from premise 3. From the contradiction $Q \wedge \sim Q$, we conclude that the conclusion $\sim R$ does indeed follow from the premises.

Exercise 3–89. Take as hypothesis:
Phyllis marries either Mountararat or Strephon. ($M \vee S$)
Phyllis does not marry Mountararat. ($\sim M$)
Prove by an indirect proof that Phyllis marries Strephon.

Exercise 3–90. Show by an indirect proof that the conclusion follows from the hypothesis:
1. Either George or Henry has $50.
2. Henry is broke.

∴* George has $50.

Exercise 3–91. Prove by an indirect proof:
1. Children eat gumdrops.
2. Gumdrops are sweet.
3. If children do not eat gumdrops, they eat red hots.
4. If gumdrops are sweet, children do not eat red hots.

∴ Children do not eat red hots.

Exercise 3–92. Prove by an indirect proof:
1. If Jack Sprat could eat no fat and his wife could eat no lean, then the platter and the pan are both licked clean.
2. The pan is dirty.
3. Mrs. Sprat eats no lean.

∴ Jack eats fat.

Exercise 3–93. Prove by an indirect proof:
1. Whenever it is raining, it is humid.
2. When it is humid, there is water in the gutter.
3. There is no water in the gutter.

∴ It is not raining.

Exercise 3–94. Prove by an indirect proof:
1. If George and Henry both have cars, then Mary and Virginia will both go to the movies.
2. Virginia stays home.
3. Henry has a car.

∴ George does not have a car.

CHECKING VALIDITY

As you have seen, even an argument that starts from false premises or nonsense premises can be correct as deduction because it follows a

---

* The symbol "∴" for "therefore" indicates the conclusion.

correct rule of deduction, like the syllogism rule. We call a deduction valid if the conclusion is implied by the hypothesis, whether the premises and conclusion are realistic or not. As an analogy, consider the legal question of acquittal versus innocence. A defendant can be acquitted according to the requirements of the law (perhaps because he could not be shown guilty beyond "all reasonable doubt") and yet not be completely innocent in the case.

You know several proof techniques based on tautologies that can be relied upon to give valid deductions: the Law of Detachment, syllogisms, contrapositives, indirect proofs. How can you be sure there are not others, or that you have applied all the ones necessary and in the right order? There is a technique that will tell you whether an argument about statements in a space is valid, that is, whether the conclusion is logically implied by the hypothesis, or whether there is a fallacy. If you are able to support each stage of the deduction by one of the tautologies you have learned (such as the Law of Detachment), then you do not need this technique; the deduction is valid.

Truth Table Test for Validity

Make a truth table showing each simple statement used in the argument. For instance, if one of the premises in the hypothesis is an implication: "If Maury is a carpenter, he can build a cupboard," carpenter-implies-cupboard, list "carpenter" and "cupboard" with their $2 \cdot 2 = 4$ truth value combinations. Next, enter in the truth table the truth value for each premise and for the conclusion. If some of these are complicated, you may want to calculate one part at a time. Then cross out every line for which one of the premises has value False. Notice that experience will often save you time on the previous stage, for a combination of simple statements that leads to a false premise can be omitted. The lines that remain are all the ones for which the hypothesis (all the premises) is True. In a valid argument, the conclusion is then True. If every entry (not crossed out) in the "conclusion" column of the truth table is True, then the deduction passes the test; it is valid. If there are any listings of False in the column, the deduction fails the test; it is *not* valid.

The truth table method is especially useful for finding holes in a faulty argument. To *dis*prove the implication from hypothesis to conclusion, all you need is one instance (line in the truth table) of True hypothesis but False conclusion.

EXAMPLE   Use a truth table to test the validity of the argument:

(*1*) $P \lor Q \Rightarrow R$ (If he wants it or needs it, he buys it.)
(2) Q (He needs it.)

∴ R (Therefore, he buys it.)

	P	Q	R	P ∨ Q	P ∨ Q ⇒ R
(1)	T	T	T	T	T
(2)	T	T	F	T	F
(3)	T	F	T	T	T
(4)	T	F	F	T	F
(5)	F	T	T	T	T
(6)	F	T	F	T	F
(7)	F	F	T	F	T
(8)	F	F	F	F	T

Cross out rows 3, 4, 7, and 8, because Q is True, from premise (2). Cross out rows 2, 4, and 6 because premise (1) is True.

Only rows 1 and 5 are left, and in those two rows the conclusion R is True. The argument is valid, as we could also show from tautologies: $Q \Rightarrow (P \lor Q) \Rightarrow R$.

Use a truth table to test the validity of the argument: **EXAMPLE**

(1) $P \lor Q \Rightarrow R$ (If he wants it or needs it, he buys it.)
(2) R (He buys it.)
(3) Q (He needs it.)

∴ P (He wants it.)

	P	Q	R	P ∨ Q	P ∨ Q ⇒ R
(1)	T	T	T	T	T
(2)	T	T	F	T	F
(3)	T	F	T	T	T
(4)	T	F	F	T	F
(5)	F	T	T	T	T
(6)	F	T	F	T	F
(7)	F	F	T	F	T
(8)	F	F	F	F	T

Cross out rows 2, 4, and 6, by premise (1). Cross out rows 2, 4, 6, and 8, by premise (2). Cross out rows 3, 4, 7, and 8, by premise (3).

In rows 1 and 5 all three premises are True, but in row 5, P is False. The argument is not valid.

Here are some famous syllogistic arguments from the works of Lewis Carroll (*Alice in Wonderland*, *Through the Looking-Glass*, and others). Try to prove the conclusion from the premises by applying tautologies you know, or by making a truth table. If you enjoy them, consult Lewis Carroll's collected works for many more, some very complicated indeed.

Exercise 3-95.

(1) Babies are illogical;
(2) Nobody is despised who can manage a crocodile;
(3) Illogical persons are despised.

Conclusion: Babies cannot manage crocodiles.

Exercise 3-96.
(1) My saucepans are the only things I have that are made of tin;
(2) I find all *your* presents very useful;
(3) None of my saucepans are of the slightest use.
Conclusion: *Your* presents to me are not made of tin.

Exercise 3-97.
(1) No potatoes of mine, that are new, have been boiled;
(2) All my potatoes in this dish are fit to eat;
(3) No unboiled potatoes of mine are fit to eat.
Conclusion: All my potatoes in this dish are old ones.

Exercise 3-98.
(1) No ducks waltz;
(2) No officers ever decline to waltz;
(3) All my poultry are ducks.
Conclusion: My poultry are not officers.

Exercise 3-99.
(1) There are no pencils of mine in this box;
(2) No sugar-plums of mine are cigars;
(3) The whole of my property, that is not in this box, consists of cigars.
Conclusion: No pencils of mine are sugar-plums.

The following arguments are not from Carroll's collection. There are some included that are not valid.

Exercise 3-100.
(1) If Joe goes to California, then Sam will go to Kansas.
(2) If Sam goes to Kansas, then George will not go to Vermont.
Conclusion: If George goes to Vermont, then Joe will not go to California.

Exercise 3-101.
(1) Mary is old.
(2) If Mary is old, she is wrinkled.
(3) Babies are wrinkled.
Conclusion: Mary is a baby.

Exercise 3-102.
(1) All single people are happy.
(2) Either George is happy or he is poor.

(3) George is not poor.
Conclusion: George is single.

Exercise 3–103.
(1) Sally has a pretty dress.
(2) If Mary goes to town, then Sally has no pretty dresses.
(3) If Mother cooks a goose for Thanksgiving dinner, Mary will have to go to town to buy potatoes.
Conclusion: Mother does not cook a goose for Thanksgiving dinner.

Exercise 3–104.
(1) If Topeka is in Kansas, then Omaha is in Nebraska.
(2) If Reno is the capital of Nevada, then Omaha is not in Nebraska.
(3) Topeka is in Kansas.
Conclusion: Reno is not the capital of Nevada.

Exercise 3–105.
(1) Gryphons are tasty.
(2) Porcupines are sharp.
(3) A good soup is both tasty and sharp.
Conclusion: Gryphon-and-porcupine soup is good.

Exercise 3–106.
(1) If the University of Oregon is in Mississippi, then the University of Wisconsin is in the Gulf of Mexico.
(2) If Miami University is in Ohio, then Oxford is in England.
(3) If the University of Oregon is not in Mississippi, then Oxford is not in England.
(4) Miami University is in Oxford, Ohio.
Conclusion: The University of Wisconsin is in the Gulf of Mexico.

There is a concise clear introduction to logic in *Sets, Logic,* and *Axiomatic Theories*, by Robert R. Stoll, W. H. Freeman and Co., San Francisco, 1961.

REVIEW OF CHAPTER 3

1. **Universe of discourse,** or **space**—all objects connected with the discussion. Let the space U be our solar system, that is, our sun and its surrounding planets and lesser bodies. Is the North (Pole) star in U? Is our moon in U? Draw a diagram of U. Figure 3–21 shows enough of our solar system for the following exercises.

2. **Statement in** U—sentence that
   (1) concerns objects in U
   (2) has exactly one truth value T or F for each object in U.
   Let $P(x)$ be

$$P(x): \text{``}x \text{ has at least one moon.''}$$

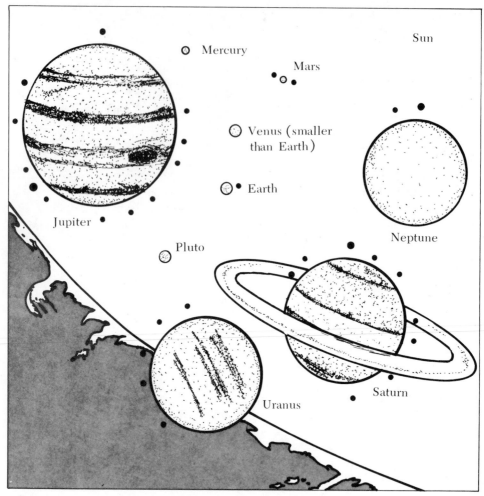

FIGURE 3–21. PLANET SIZES AND MOONS.

Show that (1) the sentence $P(x)$ can be used with each x in U; (2) the sentence $P(x)$ has truth value T for some of the objects in U, such as Jupiter, and has truth value F for all others, such as Mercury and the Earth's moon. Then $P(x)$ is a statement in U.

Show that the sentence

"x is a serious student"

is not a statement in U, because (1) it is not about objects in U.

Show that the sentence

"x is up high"

is not a statement in U, because it does not have a definite truth value for each x in U, so that requirement (2) for a statement is not met.

Draw a diagram for U and P. Show where in your diagram the x's lie that make $P(x)$ True.

3. **Not.**

P	~P
T	F
F	T

Write $\sim P(x)$ for the statement

$$P(x): \text{"}x \text{ has at least one moon."}$$

Draw a diagram showing U, P, and $\sim P$.

4. **And.**

P	Q	P ∧ Q
T	T	T
T	F	F
F	T	F
F	F	F

Let $Q(x)$ be the statement in U

$$Q(x): \text{"}x \text{ is larger than Earth."}$$

Write $P(x) \wedge Q(x)$. What is the truth value of $P(x) \wedge Q(x)$ for the object $x =$ Jupiter? What is the truth value of $P(x) \wedge Q(x)$ for $x =$ Earth? Which row of the truth table shows this case? Find a planet representing the fourth row. Draw a diagram showing where $P(x)$ is True, where $Q(x)$ is True, and where $P \wedge Q$ is True.

5. **Or.**

P	Q	P ∨ Q
T	T	T
T	F	T
F	T	T
F	F	F

Write $P(x) \vee Q(x)$ for the solar system example. What is the truth value for $P(x) \vee Q(x)$ for $x =$ Earth? Draw a diagram showing where $P \vee Q$ has truth value True.

6. **Conditional.**

R	Q	R → Q
T	T	T
T	F	F
F	T	T
F	F	T

Let $R(x)$ be the statement

$$R(x): \text{"}x \text{ is smaller than Saturn."}$$

Write the conditional $R(x) \rightarrow Q(x)$ and find its truth value for $x =$ Jupiter, $x =$ Neptune, $x =$ Earth, $x =$ Mercury. Draw a diagram showing where R, Q, and $R \rightarrow Q$ are True.

7. **Biconditional.** $P \leftrightarrow Q$ means that P and Q agree in truth value for each x in U.

P	Q	P ↔ Q
T	T	T
T	F	F
F	T	F
F	F	T

In the solar system example, Figure 3–21, find the truth value of the biconditional $P(x) \leftrightarrow Q(x)$ for each planet x shown.

8. **Tautology**—a statement that has constant truth value True for every combination of values for the basic statements involved. The patterns (rules, laws) of deduction are tautologies, always True independent of the specific meanings assigned in an application.

9. The **Law of the Excluded Middle** $[P \vee (\sim P)]$ and the **Law of Contradiction** $\sim [P \wedge (\sim P)]$ are tautologies. Translate them for the solar system example and show they are True throughout U.

10. The **Law of Detachment** $[(R \rightarrow Q) \wedge R] \Rightarrow Q$. Suppose you know that both $R(x) \rightarrow Q(x)$ and $R(x)$ have truth value True for $x =$ Neptune. Use the Law of Detachment to prove that Neptune is larger than Earth. More typically, the Law of Detachment $[H \wedge (H \Rightarrow C)] \Rightarrow C$ is used when H and $H \Rightarrow C$ are tautologies. In our example, $\sim Q \Rightarrow R$ is tautologically True, so for any x that is not larger than Earth $(\sim Q)$, the conclusion R follows.

11. **Contrapositive** of $R \rightarrow Q$ is $(\sim Q) \rightarrow (\sim R)$. Write $(\sim Q) \rightarrow (\sim R)$ for the solar system example, and find its truth value for $x =$ Neptune.

12. **Converse** of $R \rightarrow Q$ is $Q \rightarrow R$. Find which planets x in Figure 3-21 make $Q(x) \rightarrow R(x)$ True.

13. **Inverse** of $R \rightarrow Q$ is $(\sim R) \rightarrow (\sim Q)$. Translate the inverse for the solar system example.

14. **Definitions, necessary and sufficient conditions, if-and-only-if conditions.** Statements are equivalent in U if their biconditional is a tautology, $B \Leftrightarrow C$.

15. **Syllogism**—$[(S \rightarrow Q) \wedge (Q \rightarrow M)] \Rightarrow (S \rightarrow M)$. Translate the syllogism with statements
$S(x)$: "x is larger than Saturn" and $M(x)$: "x is larger than Mars."

16. **Indirect proof**—a proof that $H \Rightarrow C$ by establishing that $H \wedge (\sim C)$ is a contradiction (everywhere False). In the solar system example, let H be $S(x)$ and C be $Q(x)$. Show that $S(x) \wedge \sim Q(x)$ has truth value False throughout U, so that it is a contradiction. In this way give an indirect proof that $S(x) \Rightarrow Q(x)$.

17. **Truth table test for validity.** Make a truth table for the four combinations of values True and False for statements B and C. Show that the hypothesis $(B \rightarrow C) \wedge (C \rightarrow B)$ has the valid conclusion $B \leftrightarrow C$, by crossing out the rows for which $B \rightarrow C$ has value False and the rows for which $C \rightarrow B$ has value False.

In another truth table show that the hypothesis $B \rightarrow C$ does *not* justify the conclusion $C \rightarrow B$.

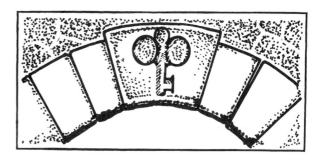

CHAPTER 3 AS A KEY TO MATHEMATICS

Formal statement logic is not a compact form of the kind of thought most popularly used, nor even of the "best" usage. It does not teach you "how to think," or "orient" you at college, or help you organize your homework assignments. It is too limited for that. However, in its very restricted sphere, or box, it does provide rules for deductive reasoning that can be written down, codified, checked mechanically.

The rules of deduction used in mathematical proofs play a role like that of a dictionary in a word game, or a code of laws in a law court. Setting aside all questions of whether the dictionary lists the most popular or best usage, the players can still agree to abide by its listings in case of dispute. Setting aside all questions of whether a code of laws conforms to modern social usage or whether the code is fair, a government can adopt the code for judging court cases. As you have seen, you can determine whether an argument follows the *pattern* of a syllogism without determining whether the premises are correct—in fact, without even understanding the vocabulary!

CHAPTER 4

Themistocles replied that a man's discourse was like to a rich
Persian carpet, the beautiful figures and patterns of which can
be shown only by spreading and extending it out; when it is
contracted and folded up, they are obscure and lost.

Plutarch's Lives

Here we see the magnificent creation of George Boole, the
formal logic of Chapter 3, as but one rich vein of a deeper lode.
By following other veins we come upon an appreciation of the
extent of the lode.

BOOLEAN ALGEBRA

The Skeleton of Logic and Other Flesh It Can Wear

You have invented a "painless learning" machine and are conducting a survey before putting it on the market. You plan to interview students in various categories, as shown in Figure 4–1: In the figure, u' stands for the opposite of u within the survey space, that is, for those who are not upper-

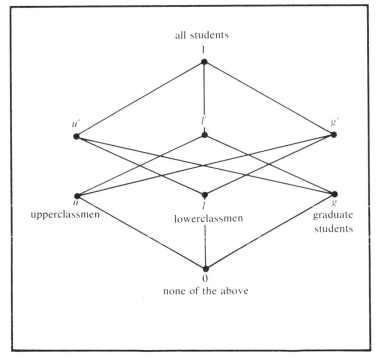

FIGURE 4–1. CATEGORIES OF STUDENTS.

classmen but are lowerclassmen or graduate students. Similarly, l' stands for non-lowerclassmen, that is, upperclassmen or graduate students, and g' stands for nongraduates, that is undergraduates. The 0 stands for none of the classifications, and the 1 for the inclusive "whole," or all classifications together.

Now we define for any two points a and b in the 8-point diagram their "join" $a \vee b$, and their "meet" $a \wedge b$: If a and b are unequal and there is a line segment connecting them, then the higher one is their join and the lower one their meet. However, if no line segment connects them directly, their join is the next point above them that can be reached along line-segment paths, and their meet is the next point below them that can be reached along line-segment paths. If $a = b$, then a, or equivalently b, serves as both $a \vee b$ and $a \wedge b$.

EXAMPLE $u \vee l = g'$, because there is no line segment directly connecting u and l and g' is the next point above both u and l that can be reached along the paths.

EXAMPLE $u \wedge 0 = 0$, because a line segment connects u and 0, so the lower one, 0, is the meet.

EXAMPLE $u \vee u' = 1$, because they are not directly connected, and 1 is the first point above them to which each is connected. Notice we have to get from u to 1 in two stages, either via l' or via g'.

You can do a kind of arithmetic in this miniature 8-point cosmos.

EXAMPLE   Find $u \vee (l \vee g)$. First, $l \vee g = u'$, so we replace the parentheses by u': $u \vee (l \vee g) = u \vee u'$. As in the example above, $u \vee u' = 1$. Then $u \vee (l \vee g) = 1$.

Exercise 4–1.   Find $u' \wedge l'$.

Exercise 4–2.   Find $u' \vee g'$.

Exercise 4–3.   Find $g \vee g'$.

Exercise 4–4.   Find $g \wedge g'$.

Exercise 4–5.   Find $u \wedge u'$.

Exercise 4–6.   Show that $b \vee 1 = 1$, no matter which of the 8 points "b" stands for.

Exercise 4-7. Show that $b \wedge 0 = 0$, no matter which of the 8 points "b" stands for.

Exercise 4-8. Show that $b \vee b' = 1$, where b takes on any of the 3 values u, l, or g.

Exercise 4-9. Show that $b \wedge b' = 0$, where b takes on any of the 3 values u, l, or g.

It turns out that the rules of this arithmetic can be packaged in compact form, as in Definition 4–1. We do not pretend that the little design shown in Figure 4–1 has much significance by itself. We use it just to introduce you through a simple case to Boolean algebras in general. After introducing the definition and showing that Figure 4–1 actually does

Definition 4–I. A **Boolean algebra** is made up of objects a, b, c, . . . , with two operations **join** $\vee$ and **meet** $\wedge$, such that the join $a \vee b$ and the meet $a \wedge b$ of any two of the objects are uniquely defined. The two operations have these properties:

For all objects a, b, c in the algebra,

1. $a \vee (b \vee c) = (a \vee b) \vee c$
1'. $a \wedge (b \wedge c) = (a \wedge b) \wedge c$ **associativity**

2. $a \vee b = b \vee a$
2'. $a \wedge b = b \wedge a$ **commutativity**

3. $a \vee (b \wedge c) = (a \vee b) \wedge (a \vee c)$
3'. $a \wedge (b \vee c) = (a \wedge b) \vee (a \wedge c)$ **distributivity**

There are objects 0 and 1 in the algebra, such that for any object b in the algebra,

4. $b \vee 0 = b$
4'. $b \wedge 1 = b$ **identity properties**

The object 0 is an **identity** for the algebra with operation $\vee$, because it leaves each object b identically the same when it "joins" b. Similarly, 1 is an **identity** for the algebra under $\wedge$, since it leaves each b unchanged when it "meets" b.

For each object b in the algebra there is an object b' for which:

5. $b \vee b' = 1$
5'. $b \wedge b' = 0$ **complementation**

illustrate a simple Boolean algebra, we are going to show that three very important systems are Boolean algebras:

a. The logic system of (true-or-false) statements and connectives in a space, as described in Chapter 3.

b. The subsets of a set with unions and intersections (which we shall define as we go along).

c. Flow systems, such as electrical circuits, trafficways, and so forth, with parallel and serial connections.

Exercise 4–10. Verify associativity of join and of meet in case a, b, c, is the triple u, l, g' in Figure 4–1.

Exercise 4–11. Show that the join and meet in Figure 4–1 have to be commutative from the way they are defined in that example.

Exercise 4–12. Verify distributivity in Figure 4–1 for the triple u, g, g'. Check both "3," distributivity of join with respect to meet, and "3'," distributivity of meet with respect to join.

Exercise 4–13. Verify that in Figure 4–1, 0 is an identity for join and 1 is an identity for meet.

Exercise 4–14. Compare Exercises 4–8 and 4–9 with the complementation properties 5 and 5'.

One typical feature of mathematics, the one that makes it applicable to such a wide variety of fields, is its abstractness. Again and again in applying mathematics, we strip away the unnecessary detail in a problem until we have only the abstract "mathematical model," making the problem appear much simplified and letting us see a familiar pattern. When you want to find the cost of 3 items at 49 cents each, you strip away temporarily the unnecessary detail, such as what items they are, which store is selling them, how much you want them, and so on, concentrating on the abstract mathematical model of 3 sets, each containing 49 objects. They might be 3 rows of 49 stars each, or 3 lecture rooms with 49 students each. Temporarily, you reduce the problem to 3×49 and so solve it simply from previous experience, without having to develop a whole arithmetic on the spot to find the total cost.

The skeleton of Definition 4–1 provides a **mathematical model** for the arithmetic of Figure 4–1. In turn, Figure 4–1 and its arithmetic provide a **realization** of the model, a fleshing-out of the skeleton.

You might compare a mathematical model to a garment pattern. Many different garments made of various materials for various purposes might be made by following the same pattern. Each would form a "realization" of the same "model."

Our next objective is to present three realizations of the Boolean algebra model, each with such far-reaching consequences that you will be amazed to see it serving as an application (of Boolean algebra) rather than as an abstraction in its own right. Our first realization is the logic system in a space introduced in Chapter 3, an idea of great generality and importance. Yet here it is just one of several examples of Boolean algebras!

A STATEMENT LOGIC AS A BOOLEAN ALGEBRA

The "objects" in a Boolean algebra are perfectly general, not described at all, so that in different realizations they can take on different characters. In the example of a logic system, the objects are the statements, in the sense of Chapter 3, that is, statements that have one of the values "true" or "false," but not both. The operations join and meet of a Boolean algebra are also left completely unspecified except that they must conform to requirements 1 through 5 and 1' through 5'. In this example, the role of "join" will be taken by the connective "or" in the usual logic sense of "either . . . or . . . or both." The role of "meet" will be taken by the connective "and." In fact, we can give an interpretation in the logic system for each of the terms in the Boolean algebra:

In the Boolean Algebra	*In the Logic System*
object $a, b, c, \ldots$	statement $a, b, c, \ldots$
join $\vee$	or $\vee$
meet $\wedge$	and $\wedge$
complement b'	negation $\sim b$ of statement b
identity under join, 0	any contradiction, such as $b \wedge \sim b$, that is, any statement that is always False
identity under meet, 1	any tautology, such as $b \vee \sim b$, that is, any statement that is always True

As a special example, take the space of counting numbers, $1, 2, 3, \ldots$, with statements "n is an even number," "n is divisible by 4," "n is divisible by 6," and their negations and connections. The complement of "n is an even number" is "n is not an even number," or, since n must still be in the space of counting numbers, "n is an odd number." The complement of "n is divisible by 4" is "n is not divisible by 4," and so on. As the identity "0," we can take "n is not a number." As "1," we can take "n is a number." (Here we mean "counting number," so that we are including all the space in "1," none of it in "0.") We can observe commutativity in statements like "n is divisible by 4 or divisible by 6" versus "n is divisible by 6 or divisible by 4," and in statements like "n is divisible by 6 and n is even" versus "n is even and it is divisible by 6." We observe associativity in statements like

"n is an even number and it is divisible by 6," [(1 $\wedge$ even) $\wedge$ div 6],

versus

"n is a number and it is even and divisible by 6," [1 $\wedge$ (even $\wedge$ div 6)].

A statement like "even $\wedge$ (div 4 $\vee$ div 6)" would mean

"n is a number that is even and is divisible either by 4 or by 6."

According to the distributive property 3′, this statement should be equivalent to

"(even ∧ div 4) ∨ (even ∧ div 6)," meaning "n is a number that is either even and divisible by 4 or even and divisible by 6."

Exercise 4–15. List all the counting numbers less than 40 that are even and are divisible by 4 or by 6. List all those less than 40 that are even and divisible by 4, and those less than 40 that are even and divisible by 6; taking these two lists together, show that they contain the same numbers as the first list.

EXAMPLE   Within the space we are considering, a statement like "n is even or n is not a number" of the form $b \lor 0$ is equivalent to "n is even," since we are restricting ourselves to numbers throughout the argument. You have seen this form in statements like "I'll capture the villain or my name isn't Sherlock Doright" or "I'll love you or water will run uphill." Each of these is of the form $b \lor 0$, the second alternative intended to be such an obvious contradiction that the first constitutes the whole of the statement.

Exercise 4–16. Write out the meaning of $b \lor 0$ when b is the statement "n is divisible by 4." According to requirement 4 of Definition 4–1, what simpler statement is equivalent to this?

Exercise 4–17. What is the form of the statement "n is not a number or n is divisible by 6"? By commutativity 3, rewrite it so as to apply the identity property 4.

EXAMPLE   In the present context, which is just another way of saying "in this space," a statement like "n is even and n is a number" is equivalent to "n is even," for all our objects of consideration are numbers. This is an illustration of the identity property 4′ of a Boolean algebra, $b \land 1 = b$.

Exercise 4–18. Write out the meaning of $b \land 1$ when b is the statement "n is divisible by 6." Compare with the statement b itself.

EXAMPLE   A statement "Each counting number is either even or odd" could be written symbolically "$b \lor {\sim}b = 1$," "the numbers that are either even or not even constitute all the numbers."

Exercise 4–19. Let b be the statement "n is divisible by 6." Write the negation ${\sim}b$, write the alternation $b \lor {\sim}b$, and check requirement 5 for complements.

Keeping the special example of counting numbers for reference, take any logic system over a space, limiting the space as we did for the counting numbers, so that we can be sure what the "1" is. We can prove that the logical connectives "or" and "and" have the Boolean algebra properties 1 through 5 and 1' through 5', using truth tables. For example, Figure 4–2 illustrates and proves property 3, distributivity of "or" with respect to "and."

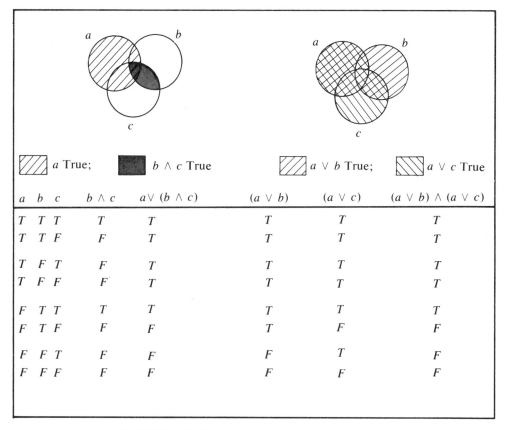

a	b	c	$b \wedge c$	$a \vee (b \wedge c)$	$(a \vee b)$	$(a \vee c)$	$(a \vee b) \wedge (a \vee c)$
T	T	T	T	T	T	T	T
T	T	F	F	T	T	T	T
T	F	T	F	T	T	T	T
T	F	F	F	T	T	T	T
F	T	T	T	T	T	T	T
F	T	F	F	F	T	F	F
F	F	T	F	F	F	T	F
F	F	F	F	F	F	F	F

FIGURE 4–2. PROOF OF PROPERTY 3.

The truth table shows that $a \vee (b \wedge c)$ and $[(a \vee b) \wedge (a \vee c)]$ have the same truth value for all combinations of a, b, c. The two compound statements are equivalent.

Exercise 4–20. In the space of counting numbers let

a be "n is an even number"

b be "n is divisible by 4"

c be "n is divisible by 6."

Interpret property 3, distributivity of "or" with respect to "and," for this case.

Exercise 4–21. Continuing the notation of Exercise 4–20, translate property 3′, distributivity of "and" with respect to "or." Name a number for which *a* is True but *b* ∨ *c* False.

Exercise 4–22. Use truth tables to show that property 3′ holds for a logic system.

Because the logic system over any assigned space is a realization of the Boolean algebra model, any theorems that can be proved for the Boolean algebras can be used immediately without more proof for any logic system. By combining the objects with join and meet according to the properties 1 through 5 and 1′ through 5′, you can prove abstract theorems about Boolean algebras. For instance,

$$a \vee (a' \wedge b) = (a \vee a') \wedge (a \vee b), \text{ by property 3}$$
$$= 1 \wedge (a \vee b), \text{ by property 5}$$
$$= (a \vee b) \wedge 1, \text{ by property 2'}$$
$$= a \vee b, \text{ by property 4'.}$$

Now apply this little theorem to the space of people invited to a party, where each guest is either *a* = male or else is female and *b* = escorted. Since $a \vee (a' \wedge b) = a \vee b$, we can say immediately that each guest is male or escorted.

This short example barely suggests the richness of possible applications. As you will see, other realizations of the mathematical model open whole classes of further applications.

AN ALGEBRA OF SETS AS A BOOLEAN ALGEBRA

The symbols ∨ and ∧ may already have suggested to you the symbols ∪ for "union" and ∩ for "intersection" used in combining sets. We state in the form of a definition the essentials of an algebra of sets:

> **Definition 4–2.** Let *U* be a nonempty set called the **space.** Let *S, T, R,* and so forth be sets in the space *U*. We leave the word "set" itself undefined to avoid circularity, but we do assume that each set has a **discriminating property** that determines for each object *x* in the space *U* whether it is in a given set *S* or not. We say two sets are **equal**, written *S* = *T*, if they have exactly the same members. The **union** of two sets *S* and *T*, written *S* ∪ *T*, is the set in *U* made up of all objects in *U* that are in *S* or in *T* or in both. The **intersection**, written *S* ∩ *T*, is the set in *U* made up of all objects in *U* that are in both *S* and *T*. The **empty set,** ∅, is the set having no members in it. The **complement** $\bar{S}$ of a set is the set of objects in *U* that are not members of *S*.

Let the space U for this example be your class in mathematics. Let **EXAMPLE**
S be the set of class members who are male, which we can write conveniently in symbols:

$S = \{x \mid x$ is male$\}$, read "S is the set of class members x such that
$\qquad\qquad\qquad\qquad\qquad\qquad x$ is male."

Let T be those whose last names start with letters from A through H,

$$T = \{x \mid x \text{ has initial } A\text{--}H\}.$$

Let R be those whose last names start with letters I through P,

$$R = \{x \mid x \text{ has initial } I\text{--}P\}.$$

Let

$$F = \{x \mid x \text{ has initial } Q\text{--}W\},$$

and let

$$G = \{x \mid x \text{ has initial } X, Y, \text{ or } Z\}.$$

Each of these sets has a discriminating property for determining which class members belong to the set.

The set $T \cup R$ in this case is made up of the class members whose last names start with letters A through P:

$$T \cup R = \{x \mid x \text{ has initial } A\text{--}H\} \cup \{x \mid x \text{ has initial } I\text{--}P\}$$
$$= \{x \mid x \text{ has initial } A\text{--}P\}.$$

The set $S \cap T$ is made up of the male class members with initials A through H:

$$S \cap T = \{x \mid x \text{ is male}\} \cap \{x \mid x \text{ has initial } A\text{--}H\}$$
$$= \{x \mid x \text{ is male and has initial } A\text{--}H\}.$$

It may be that there are no people in the class with initial X, Y, or Z, so that the set G is empty. In that case we say $G = \varnothing$, since G and $\varnothing$ have exactly the same members (none at all!).

The complement $\bar{S}$ of the set S is made up of all the females in the class.

$$\bar{S} = \{x \mid x \text{ is not male}\} = \{x \mid x \text{ is female}\}.$$

Exercise 4–23. In Figure 4–3 find the sets

$$C = \{\text{boxes} \mid \text{box is closed}\}$$
$$T = \{\text{boxes} \mid \text{box is tied}\}$$

Which boxes are in the set $C \cap T$? the set $T \cap C$? Which boxes are in the set $\bar{C} \cup T$? the set $C \cup \bar{T}$?

Exercise 4–24. Continuing Exercise 4–23, let R be the set of boxes in Figure 4–3 that are rounded in cross section. Find $R \cap (C \cup T)$ and $(R \cap C) \cup (R \cap T)$.

Exercise 4–25. Continuing the last two exercises, find $C \cup (R \cap T)$ and $(C \cup R) \cap (C \cup T)$.

FIGURE 4–3.

Exercise 4–26. In the photograph of musical instruments, Figure 4–4, find the set S of stringed instruments, the set W of wind instruments, the set $\bar{S} \cap \bar{W}$, and the set M of instruments made of metal. Find $M \cap S$ and $M \cap W$.

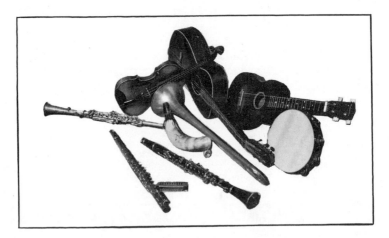

FIGURE 4–4. MUSICAL INSTRUMENTS.

Exercise 4–27. Let the space U be the numbers $-10, -9, -6, -5, -4, -1, 0, 1, 2, 4, 8, 12, 16$. In this space, the set $S = \{x \mid x \text{ is even}\}$ can

be given by listing its members: $S = \{-10, -6, -4, 0, 2, 4, 8, 12, 16\}$.
Give the following sets of the space by listing their members:

$A = \{x \mid x$ is divisible by 3$\}$.

$B = \{x \mid x$ is negative$\}$.

$C = \{x \mid x$ is positive$\}$.

$D = \{x \mid x$ is divisible by 6$\}$.

$\bar{S} = $ complement of S.

$B \cup D$.

$B \cap D$.

$S \cap A$. Compare with D.

$\overline{B \cup C}$, the complement of the union $B \cup C$. Compare with $\emptyset$.

$B \cap C$. Compare with $\emptyset$.

$S \cup \bar{S}$. Compare with U, the whole space.

Exercise 4–28. Let the space U be a class of students. Let the set

$A = \{x \mid x$ does well on examinations$\}$

$B = \{x \mid x$ does well on class recitation$\}$

$C = \{x \mid x$ does well on homework$\}$.

Find $A \cup B$.

Find $B \cup C$.

Find $A \cap (B \cup C)$.

Find $A \cap B$.

Find $A \cap C$.

Find $(A \cap B) \cup (A \cap C)$ and compare with $A \cap (B \cup C)$.

Exercise 4–29. Let U be the space of gasoline-powered cars. For definiteness, in fact, restrict U to cars that are signed up for a particular gasoline economy run. Let $W = \{x \mid x$ gets up to 35 miles per gallon$\}$. To make this a properly defined set, assume the mileage is computed in some specified way.
Let

$$X = \{x \mid x \text{ gets up to 70 miles per gallon}\}$$
$$Y = \{x \mid x \text{ gets over 10 miles per gallon}\}$$
$$Z = \{x \mid x \text{ gets between 20 and 30 miles per gallon}\}.$$

Describe $X \cup Z$, $W \cap X$, $W \cap Y$, $W \cap Z$, $Y \cup Z$, and $\bar{W}$.

Exercise 4–30. Let U be percentage scores from 0 to 100 in a college course.
Let

$$A = \{x \mid 90 \leq x \leq 100\}$$
$$B = \{x \mid 80 \leq x < 90\}$$

$$C = \{x \mid 70 \leq x < 80\}$$
$$D = \{x \mid 60 \leq x < 70\}$$
$$F = \{x \mid x < 60\}.$$

What is the intersection of any two different sets A, B, C, D, or F? Use set notation to describe the acceptable scores for graduate credit, above "C"; acceptable scores for passing the course, above "F"; and acceptable scores for admission to the next course in the sequence, above "D."

The sets in a space with their algebra of union, intersection, and complementation form an example of a Boolean algebra. To show this, we need to show that the set algebra has properties 1 through 5 and 1' through 5' of Definition 4–1. The properties of associativity and commutativity for both union and intersection are familiar from our experience with set membership. To check a more complex property like distributivity, it is helpful to use a "Venn" diagram, similar to the diagrams used in Chapter 3 to show where a statement is True. For example, return to Figure 4–2 and and now let the circle a be taken to enclose all the members of a set a, circle b enclose set b, and circle c enclose set c. We show that the sets $a \cup (b \cap c)$ and $(a \cup b) \cap (a \cup c)$ have exactly the same members and so are equal. For diagrams of sets, we often draw a rectangle to enclose the whole space U. This enables us to represent the complement of a set as all points of the rectangle U that are not within the set circle. Figure 4–5 illustrates some sets in diagram form.

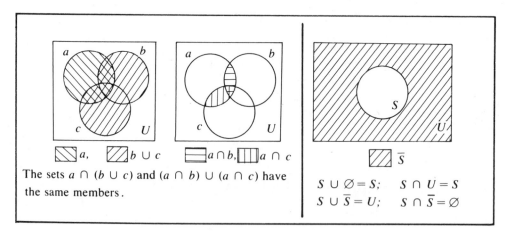

The sets $a \cap (b \cup c)$ and $(a \cap b) \cup (a \cap c)$ have the same members.

$S \cup \varnothing = S;$ $\quad S \cap U = S$
$S \cup \bar{S} = U;$ $\quad S \cap \bar{S} = \varnothing$

FIGURE 4–5. SOME SET DIAGRAMS.

Once we had shown that a logic system over an assigned space is a realization of a Boolean algebra, we tried applying the theorem

$$a \lor (a' \land b) = a \lor b$$

about Boolean algebras to a specific logic system. If we rewrite this same theorem in the form of algebra of sets, it reads

$$S \cup (\bar{S} \cap T) = S \cup T.$$

Figure 4–6 illustrates the way this theorem applies to sets.

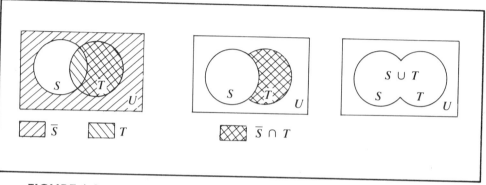

FIGURE 4–6. A THEOREM FROM BOOLEAN ALGEBRA APPLIED TO SETS.

All we need to do to get free theorems about sets is to reinterpret theorems about Boolean algebras. Now, though, we leave sets and introduce a third realization of the Boolean model.

A SWITCHING CIRCUIT AS A BOOLEAN ALGEBRA

In this realization, the objects $a, b, c, \ldots$ are switching circuits, each switch having two settings, open and closed. A join $a \lor b$ is represented as two switching circuits a and b connected in parallel. A meet $a \land b$ is represented as two switching circuits connected in series. To represent a complement a' we use a "double pole" switch that closes a when it opens a' and opens a when it closes a'. Figure 4–7 illustrates such circuits. Notice that there are two switches labeled "a" in the circuit for $(a \land b) \lor (a \land c)$. You can think of these two "a" switches as connected to form a "double throw" switch that opens both together or closes both together.

When are two circuits considered equal? We will call them equal if they have the same net effect under all conditions; that is, if they both yield an open circuit or both yield a closed circuit for any combination of switch settings. For example, in Figure 4–8 the circuit for $a \lor (b \land c)$ equals the circuit for $(a \lor b) \land (a \lor c)$, because as shown in the accompany-

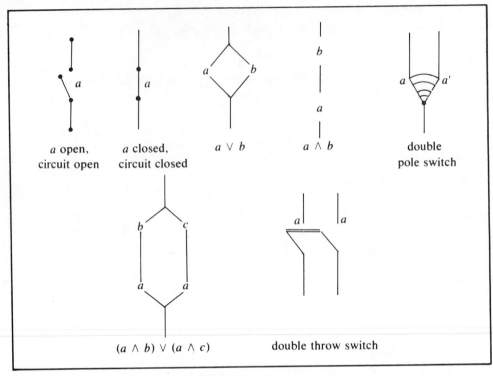

FIGURE 4–7. COMBINATIONS IN CIRCUITS.

ing table, the circuits are both open or both closed for all 8 possible settings of the three switches involved, *a*, *b*, and *c*.

Exercise 4–31. Show which two of the circuits in Figure 4–9 are equal.

Exercise 4–32. Verify that switching circuits exhibit properties 1, 1′, 2, and 2′.

Exercise 4–33. Figure 4–8 shows that switching circuits have property 3′. Show that they also have property 3′. Recall the double throw switch in Figure 4–7.

For switching circuits we can define the identity 0 to be a circuit with a break in it, one that is always open. The identity 1 can be represented by a short-circuit, a circuit that is always closed. Figure 4–10 illustrates properties 4, 4′, 5, and 5′.

Once again, then, you can take theorems freely from the Boolean algebra supply, and with the agreed-upon substitutions or interpretations write them as theorems about switching circuits.

$a \lor (b \land c)$

$(a \lor b) \land (a \lor c)$

setting of switch			circuit for	circuit for
a	b	c	$a \lor (b \land c)$	$(a \lor b) \land (a \lor c)$
Open	Open	Open	Open	Open
O	O	Closed	O	O
O	C	O	O	O
O	C	C	C	C
C	O	O	C	C
C	O	C	C	C
C	C	O	C	C
C	C	C	C	C

FIGURE 4–8. TWO EQUIVALENT CIRCUITS.

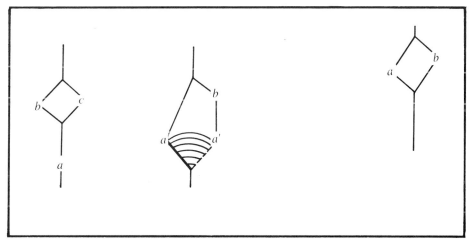

FIGURE 4–9.

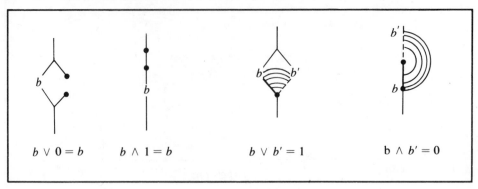

$$b \vee 0 = b \qquad b \wedge 1 = b \qquad b \vee b' = 1 \qquad b \wedge b' = 0$$

FIGURE 4–10.

THE PRINCIPLE OF DUALITY

The position at present is illustrated in Figure 4–11. We have an abstract mathematical model with three different realizations. Theorems

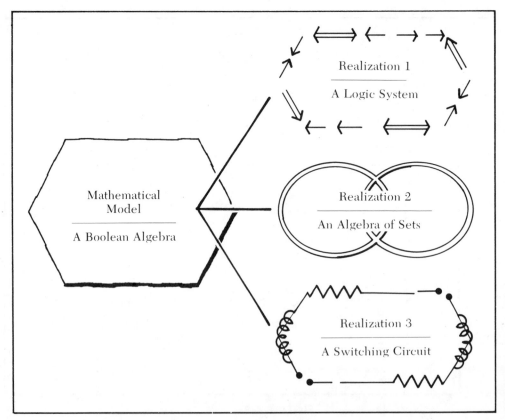

Mathematical
Model
―――――――
A Boolean Algebra

Realization 1
―――――――
A Logic System

Realization 2
―――――――
An Algebra of Sets

Realization 3
―――――――
A Switching Circuit

FIGURE 4–11.

about the model can be interpreted as theorems about each of the realizations. However, we can carry out even more ambitious programs, too. As emphasized by the numbering 1 through 5 and 1′ through 5′, the properties of a Boolean algebra fall into **dual** pairs. For each property about joins there is a corresponding property about meets, and vice versa. The dual of each object in the algebra is its complement, with the two identities serving as a complementary pair, $0' = 1$ and $1' = 0$ (proved on page 118). This makes it possible to prove a theorem about meets for every theorem about joins, simply by interchanging their roles in every step of the proof.

You can prove as follows that the 0 of a Boolean algebra is unique, **EXAMPLE** that is, there is no other element, say z, that is also an identity with respect to join. To prove this, suppose there is another identity z with respect to join. Form the join

$$z \vee 0.$$

From property 4, $z \vee 0 = z$.
From property 2, $z \vee 0 = 0 \vee z$.
Then from property 4, using z as an identity,

$$z \vee 0 = 0 \vee z = 0.$$

Since both z and 0 equal $z \vee 0$, they are the same element of the algebra. What is the dual theorem? The dual is that the 1, the identity with respect to meet, of a Boolean algebra is unique. The proof can be adapted from the proof of uniqueness of 0 by substituting meets for joins:
From $u \wedge 1$, supposing that u is another identity with respect to meet.
From property 4′, $u \wedge 1 = u$.
From property 2′, $u \wedge 1 = 1 \wedge u$.
Then from property 4′, using u as an identity,

$$u \wedge 1 = 1 \wedge u = 1.$$

Since both u and 1 equal $u \wedge 1$, they are the same element of the algebra.

Next, we establish a useful result about Boolean algebras:

Theorem 4–1. The complement of an element b in a Boolean algebra is unique.

PROOF. Suppose an element b has complements b' and b^*. We can prove that b' and b^* must be equal, that is, they must be two names for the same unique complement.

$$
\begin{aligned}
b' &= b' \wedge 1, \text{ by } 4', \\
&= b' \wedge (b \vee b^*), \text{ by } 5, \\
&= (b' \wedge b) \vee (b' \wedge b^*), \text{ by } 3', \\
&= 0 \vee (b' \wedge b^*), \text{ by } 2' \text{ and } 5', \\
&= b' \wedge b^*, \text{ by } 2 \text{ and } 4.
\end{aligned}
$$

In exactly the same way, we can show that $b*$ is also equal to $b* \wedge b'$, which by property 2' equals $b' \wedge b*$. Then $b' = b*$. ∎

Exercise 4–34. We can prove that the complement of the identity 0 in a Boolean algebra is the identity 1, as follows:

$$0 \vee 1 = 1 \vee 0, \text{ by property 2,}$$
$$= 1, \text{ by property 4,}$$

so

$$0 \vee 1 = 1.$$
$$0 \wedge 1 = 0, \text{ by property 4'.}$$

Then 1 exhibits properties 5 and 5' of a complement 0' for the identity 0. Since, by Theorem 4–1 the complement for each element is unique, 1 is the complement 0'.

Apply the Principle of Duality to prove the dual theorem, $1' = 0$.

Exercise 4–35. Supply the property necessary to justify each step in the following proof.

For each element b in a Boolean algebra, $b \vee b = b$.

Proof: $b \vee b = (b \vee b) \wedge 1$, by property __,
$$= (b \vee b) \wedge (b \vee b'), \text{ by property } \_\_,$$
$$= b \vee (b \wedge b'), \text{ by property } \_\_ \text{ applied in reverse,}$$
$$= b \vee 0, \text{ by property } \_\_,$$
$$= b, \text{ by property } \_\_.$$

Exercise 4–36. Write the dual theorem to that in Exercise 4–35.

Exercise 4–37. Prove the theorem stated in Exercise 4–36 by writing the dual of each step in the proof of Exercise 4–35.

Exercise 4–38. You can prove that for any element b in a Boolean algebra, $b \vee 1 = 1$, for

$$b \vee 1 = b \vee (b \vee b'), \text{ by property } \_\_,$$
$$= (b \vee b) \vee b', \text{ by property } \_\_,$$
$$= b \vee b', \text{ as proved in Exercise 4–35,}$$
$$= 1, \text{ by property } \_\_.$$

Draw a switching circuit illustrating this theorem.

Exercise 4–39. State and prove the dual to the theorem in Exercise 4–38. Apply the theorem to sets.

The De Morgan Laws

The following theorem about Boolean algebras gives us "laws" for finding the complement of a join or of a meet.

Theorem 4–2. If a and b are elements in a Boolean algebra, then $(a \vee b)' = a' \wedge b'$ and $(a \wedge b)' = a' \vee b'$ (the first and second De Morgan Laws).

PROOF. As you have seen, the Principle of Duality makes the proof of the second part immediate if we can prove the first part, so we shall prove only that the complement of $a \vee b$ is $a' \wedge b'$. To prove that these elements are complements, we have to establish both property 5 and property 5', so the proof breaks into two sections. First we form the join of $a \vee b$ and $a' \wedge b'$ and show that the result is 1, as required in property 5.

$$(a \vee b) \vee (a' \wedge b') = [(a \vee b) \vee a'] \wedge [(a \vee b) \vee b'], \text{ by property } \_,$$
$$= [(b \vee a) \vee a'] \wedge [a \vee (b \vee b')], \text{ by } \_,$$
$$= [b \vee (a \vee a')] \wedge [a \vee 1], \text{ by } \_,$$
$$= (b \vee 1) \wedge (a \vee 1), \text{ by } \_,$$
$$= 1 \wedge 1, \text{ by the theorem in Exercise 4–38,}$$
$$= 1, \text{ by property } \_.$$

Exercise 4–40. Supply the properties used in the proof just above.

Exercise 4–41. Complete the second section of the proof by showing that $(a \vee b) \wedge (a' \wedge b') = 0$.

Finally we appeal to Theorem 4–1 to say that since $a' \wedge b'$ has been shown to be a complement for $a \vee b$, it is the unique complement, $(a \vee b)'$. The other De Morgan Law follows from duality. ∎

Exercise 4–42. Interpret the first De Morgan Law for a logic system.

Exercise 4–43. Interpret the first De Morgan Law for sets.

Exercise 4–44. Interpret the first De Morgan Law for a switching circuit.

Exercise 4–45. Interpret the second De Morgan Law for a logic system.

Exercise 4–46. Interpret the second De Morgan Law for sets.

Exercise 4–47. Interpret the second De Morgan Law for a switching circuit.

PARTIAL ORDERING

You have applied theorems about Boolean algebras to logic systems, to sets, and to circuits. Then you have seen that Boolean algebras have a duality principle, and that the whole principle can be applied to any realization of a Boolean algebra. Yet, rich as these sources of theorems are, they do not exhaust the possibilities.

As you worked with algebras of sets, you noticed the similarity of set diagrams to the diagrams of logic systems. Then to analyze a switching circuit you made a table very much like a truth table for a logic system, with "open" and "closed" replacing "false" and "true." These comparisons suggest that a useful feature of one of the realizations may carry over to the others.

One useful feature of set algebras is the idea of a subset. Let U be a space and $S, T, R, \ldots$ sets in U. It may happen that every element of one of the sets, say S, is also an element of another set, say T. Then we say that S is a **subset** of T, or in symbols

$$S \subseteq T.$$

In this case, $S \cap T = S$, $S \cup T = T$, $S \cap \bar{T} = \varnothing$, and $\bar{S} \cup T = U$. (See Figure 4–12.)

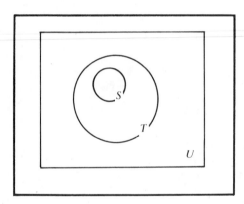

FIGURE 4–12. Subset.

EXAMPLE   Let U be the space of United States historical chronology. If S is the set of historical events that occurred during the Jefferson administration and T is the set of events prior to the Civil War, then $S \subseteq T$. In fact, since there were events in T that did not occur during the Jefferson administration, we can write $S \subset T$, or "S is a **proper subset** of T." In this example

$S \cap T = \{$events $x \mid x$ occurred during the Jefferson administration and prior to the Civil War$\}$

$\qquad = \{x \mid x$ during Jefferson's administration$\}$

$\qquad = S.$

$S \cup T = \{x \mid x$ during Jefferson's administration or prior to Civil War$\}$

$\qquad = \{x \mid x$ prior to Civil War$\}$

$\qquad = T.$

$S \cap \bar{T} = \{x \mid x$ during Jefferson's administration and during or after
$$\text{Civil War}\}$$
$$= \varnothing, \text{ the empty set.}$$

$\bar{S} \cup T = \{x \mid x$ not during Jefferson's administration or prior to Civil War$\}$
$$= U, \text{ the whole space.}$$

Now let R be the events from halfway through the Jefferson administration to World War I. Then $S \nsubseteq R$, because the first half of Jefferson's administration is not in R; $R \nsubseteq S$, because the more recent events in R are not in S; $R \nsubseteq T$, because the recent events in R occurred after the Civil War; $T \nsubseteq R$, because there were events prior to the Civil War before Jefferson took office.

Exercise 4–48. In Figure 4–3 show that $T \subset C$ for the set T of tied boxes and the set C of closed boxes. Show that $\bar{C} \subset \bar{T}$.

Exercise 4–49. Let U be the numbers 1, 2, 3, 4, 5. Let $S = \{1, 2\}$, $T = \{1, 3\}$, and $R = \{1, 2, 4\}$. Show that $S \subseteq R$, but that no set inclusion exists between T and S nor between T and R. Find $S \cup R$, $S \cap R$, $S \cap \bar{R}$, and $\bar{S} \cup R$.

Exercise 4–50. Show by set diagrams that $S \cap T \subseteq S$, that $S \cap T \subseteq T$, that $S \subseteq S \cup T$, and that $T \subseteq S \cup T$.

Exercise 4–51. Let U be the students in a class, let S be the set of those who are wearing a red garment, and let T be those who are under 5 feet 6 inches tall. Show that if R is the set of those under 5 feet tall dressed in red, then $R \subseteq S$ and $R \subseteq T$. Find $R \cup \bar{S}$ and $\bar{R} \cap T$.

The sets in U are said to be partially ordered by set inclusion $\subseteq$. The ordering is "partial," for overlapping sets cannot be ordered if neither is completely contained in the other. However, some special pairs can always be ordered. Each set S has itself as a subset, $S \subseteq S$, and if $S \subseteq T$ and $T \subseteq R$, then $S \subseteq R$.

Next we define a partial ordering for the Boolean algebra model, taking our cue from set inclusion. Then we can see what happens if we interpret the partial ordering for the other realizations, logic systems and circuits.

Definition 4–3. For elements a, b of a Boolean algebra, $a \leq b$ (read "a is less than or equal to b") means $a \vee b = b$.

It can be proved that the four conditions

$$a \vee b = b, \text{ that is, } a \leq b,$$

$$a \wedge b = a,$$
$$a \wedge b' = 0,$$

and

$$a' \vee b = 1$$

are all equivalent. Since each one holds if and only if the other three hold, any of them could be used to define the same partial ordering as Definition 4–3.

For every element b in the algebra,

$$0 \leq b \leq 1,$$

so that 0 is the "least" element and 1 is the "greatest" element in the algebra:

Exercise 4–52. Prove directly from Definition 4–3 that $0 \leq b$ for every element b in a Boolean algebra.

Exercise 4–53. Prove directly from Definition 4–3 that $b \leq 1$ for every element b in a Boolean algebra.

Exercise 4–54. Prove that $b \leq b$ for every element b in a Boolean algebra.

Theorem 4–3. In a Boolean algebra, if $a \leq b$ and $b \leq c$, then $a \leq c$.

PROOF. We use Definition 4–3 to write $a \leq b$ as $a \vee b = b$ and $b \leq c$ as $b \vee c = c$. Then

$$c = b \vee c = (a \vee b) \vee c,$$
$$= a \vee (b \vee c), \text{ by property 1,}$$
$$= a \vee c.$$

Then $a \vee c = c$, so $a \leq c$. ∎

How can we interpret this partial ordering in a statement logic? We shall use the condition for $a \leq b$ in the form $a' \vee b = 1$ rather than the form $a \vee b = b$, as the former is easier to recognize in statement logic. Suppose P and Q are statements bearing the logical relation

"$\sim P \vee Q$ is a tautology (everywhere True)"

This is equivalent to the tautology

"$P \Rightarrow Q$ (P logically implies Q, or $P \Rightarrow Q$ tautologically True)."

Formal tautological implication is a partial ordering in a statement logic. We cannot always say of every pair of statements that one of them formally implies the other. That is why the ordering is "partial." However, we do have for any statements P and Q, any contradiction 0, and any tautology 1, the implications $P \Rightarrow P$, $0 \Rightarrow P$, $P \Rightarrow 1$, $P \Rightarrow (P \lor Q)$, and $(P \land Q) \Rightarrow P$. (See Figure 4–13.) All these implications are special forms of theorems about partial ordering in any Boolean algebra. The analog for statement logic of Theorem 4–3 is

"If $P \Rightarrow Q$ and $Q \Rightarrow R$, then $P \Rightarrow R$": the Law of Syllogisms.

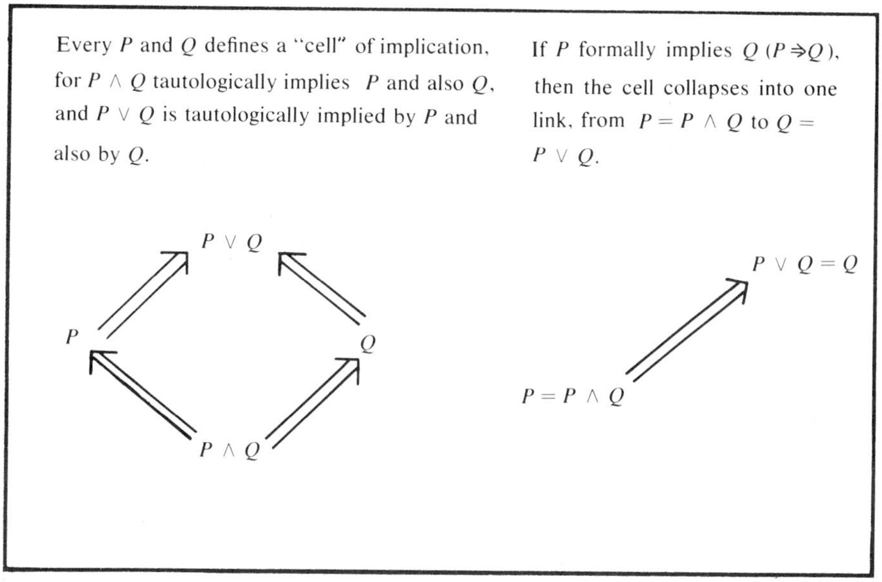

Every P and Q defines a "cell" of implication, for $P \land Q$ tautologically implies P and also Q, and $P \lor Q$ is tautologically implied by P and also by Q.

If P formally implies Q $(P \Rightarrow Q)$, then the cell collapses into one link, from $P = P \land Q$ to $Q = P \lor Q$.

FIGURE 4–13. Diagram of statements P and Q without and with a formal implication between them.*

When we describe a set by the notation

$$S = \{x \mid x \text{ has property } P\}$$

we are close to a very close link between sets in a space and statements in a statement logic. All we need to do is let the statement P read "x has the property P." Then an x that gives the *statement* P the truth value "True" is an element of the *set* S; and an x that gives the statement P the truth value False is not an element of the set S (or equivalently, an x that makes $\sim P$ True is in $\bar{S}$).

Exercise 4–55. In Figure 4–14 show that the set F of "foreground people marked with an F" is contained in the set S of seated people. Let R be the set of people in the right half of the picture. Show that no set inclusion holds between R and S or between F and S.

---

* We are grateful to reviewer Herbert J. Nichol for suggesting this diagram.

FIGURE 4–14. Picnickers.

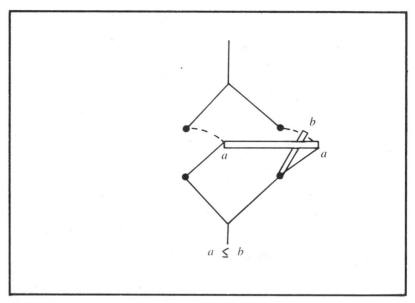

FIGURE 4–15. When a is open, b can be either open or closed, but closing a automatically closes b.

Exercise 4–56. In Figure 4–14 let $F(x)$ stand for the statement "x is in the foreground, marked with an F," and let $S(x)$ stand for "x is seated." Show that the conditional $F(x) \rightarrow S(x)$ has truth value True for each person in the picture. Let $R(x)$ stand for "x is in the right half of the picture." Find a person for whom the conditional $R(x) \rightarrow S(x)$ is False. Find someone for whom $S(x) \rightarrow R(x)$ is False. Show that $F(x) \rightarrow R(x)$ and $R(x) \rightarrow F(x)$ are sometimes False.

Implication as a partial ordering in a statement logic suggests how to interpret $a \leq b$ for switching circuits: Whenever circuit a is closed, circuit b is closed (see Figure 4–15). This interpretation suggests that we could go directly from a logic system to an equivalent switching circuit and so build a "reasoning" machine. You might try to build such a circuit for a syllogism, "If P implies Q and Q implies R, then P implies R". If you do not have the materials for an electrical circuit, a finger maze, a drawing you trace with your finger instead of with electric current, will do just as well, or you might make an impressive display with marbles rolling through channels having gates a, b, c, and so forth.

A readable explanation of Boolean algebras and their relation to logic, sets, and switching circuits appears in Chapter 3 of *The Nature of Mathematics*, by Frederick H. Young, published by John Wiley & Sons, 1968.

REVIEW OF CHAPTER 4

1. Let the space U be the objects in a flower garden. Let $A(x)$, $V(x)$, and $M(x)$ be the statements

$A(x)$: "x is animal" (that is, derived from the animal kingdom for instance a honeybee),

$V(x)$: "x is vegetable" (grass, for instance),

$M(x)$: "x is mineral" (part of the soil, for instance).

Using this example, illustrate features 1, 1′, 2, 2′, 3, and 3′ of a Boolean algebra. Memorize the names of the three properties for each operation.

2. In the garden example, find a statement that can play the role of the identity O in the Boolean algebra. Find a statement for the identity 1.

3. What is the negation of $A(x)$ in the garden?

4. Let the set A of elements in the garden be defined by

$$A = \{x \mid x \text{ is animal}\}.$$

Define V and M analogously. Draw a Venn diagram for $A \cup V$ and for $V \cap M$. Suppose one soil sample from the garden contains mineral phosphate and vegetable humus. In what set does the sample lie?

5. Find the identity sets in the garden example.

6. What is the complementary set for the set V?

7. What is the set $\overline{A \cup V}$? Does it equal the set $\bar{A} \cap \bar{V}$?

8. A pinball machine is arranged so that to pass through a $100 gate a marble must first pass through a $5 gate. Compare this situation to an implication between statements

$P(x)$: "marble x passes a $5 gate"

$Q(x)$: "marble x passes a $100 gate."

9. Let the set P be made up of all marbles (see question 8) that pass a $5 gate and let the set Q be made up of all marbles that pass a $100 gate. Express the implication of question 8 in subset terminology.

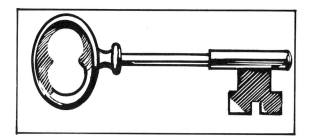

CHAPTER 4 AS A KEY TO MATHEMATICS

You could go on for the rest of the semester proving theorems about Boolean algebras by applying the rules of their arithmetic, properties 1 through 5 and 1′ through 5′. You could apply each theorem to the three realizations, interpreting the result appropriately. Of course, the spaces that can serve as examples of logic systems or sets or circuits are endlessly varied, and some of them have considerable importance of their own.

Yet what you need from this chapter to improve your understanding of mathematics is none of these things, but something much easier. You need only appreciate the basic idea of a mathematical model abstracted from experience, yielding theorems about the model that can then be applied to various realizations. The abstractness of mathematics is an important part of its character. We shall not try to define mathematics, but certainly the use of abstraction enters prominently.

CHAPTER 5

"It is nothing," said Poirot modestly. "Order, method, being prepared for eventualities beforehand—that is all there is to it."

Hercule Poirot, in
The Mystery of the Blue Train, *by Agatha Christie*

Matrices illustrate the amazing power of well-chosen notation, symbols, and form. Sometimes the mere adoption of an especially appropriate tabulation scheme can simplify a problem. The reduction of a technique to a systematic routine makes it possible to turn problems over to less skilled personnel or to computers.

MATRICES

Tabulating Information

You are helping with arrangements for a statewide music conference for junior high school bands and glee clubs. Most groups have a few adults with them, teachers, parents, and other advisors, whom you register separately, since you plan adult activities for them. You develop a systematic way of tabulating registration blanks as they are mailed in from the various junior high schools around the state. You make out a 2 by 2 table like that in Figure 5–1 for each school.

MIDTOWN JUNIOR HIGH SCHOOL		
	students	adults
band	23	2
glee club	35	6

FIGURE 5–I. 2 BY 2 TABLE FOR MIDTOWN.

Exercise 5–1. The registration blank from Newton lists 34 band students with 2 teachers and 4 parents, 15 girls' glee club students with 1 teacher and 3 parents, and 17 boys' glee club students with 1 parent and 2 adult counselors. Show how to fill out a 2 by 2 table like the one for Midtown in Figure 5–1, tabulating the information from Newton.

Exercise 5–2. Olesburg registers 8 band students with no adults, 5 girl glee club members and 2 boys in the glee club with 1 teacher and 1 parent. Show this information in a 2 by 2 table like that in Figure 5–1.

Exercise 5–3. Find the total number of band student registrants from Midtown, Newton, and Olesburg. (See Figure 5–1 and Exercises 5–1 and 5–2.)

Exercise 5–4. Find the total number of adults supervising band students for Midtown, Newton, and Olesburg. (See Figure 5–1 and Exercises 5–1 and 5–2.)

Exercise 5–5. You want to send a certain information sheet to each glee club registrant, student, or adult. What is the total number of sheets you need for Midtown, Newton, and Olesburg? (See Exercises 5–1 and 5–2 and Figure 5–1.)

Exercise 5–6. You hope to arrange invitations for all the adults to stay overnight in private homes. How many will this be for Midtown, Newton, and Olesburg? (See Exercises 5–1 and 5–2 and Figure 5–1.)

To plan for the Music Conference Picnic, you allow 2 hot dogs for each student, 1 for each adult, and $\frac{1}{4}$ cup of potato salad for each student, $\frac{1}{2}$ cup for each adult. Figure 5–2 shows how you tabulate the total registration and the picnic information.

	students	adults		hot dogs	potato salad
band	197	32	student	2	$\frac{1}{4}$
glee club	258	49	adult	1	$\frac{1}{2}$

FIGURE 5–2. TABLES.

The State Band Society and the State Association of Glee Clubs are billed separately, so you keep track of them separately. To find how many hot dogs to order for the Band Society, you calculate 197·2, because each student gets 2 hot dogs, and 32·1, because each adult gets one hot dog. Then the bands need

$$197 \cdot 2 + 32 \cdot 1 = 426 \text{ hot dogs.}$$

Exercise 5–7. How many hot dogs do the glee clubs need?

Exercise 5–8. How much potato salad do the bands need?

Exercise 5–9. How much potato salad do the glee clubs need?

Now put yourself in a new situation. You work for a florist. Two of your most popular floral arrangements are a centerpiece for a table and an arrangement to decorate a mantelpiece, both using carnations and snapdragons, as shown in Figure 5–3.

	carnations	snapdragons
table	9	4
mantel	3	8

FIGURE 5–3. INGREDIENTS OF FLORAL ARRANGEMENTS.

How many carnations are used in the mantelpiece decoration accord- **EXAMPLE** ing to Figure 5–3? In the second (horizontal) row and the first (vertical) column is the number of carnations used, 3.

Exercise 5–10. How many snapdragons are used in the centerpiece for a table?

Exercise 5–11. How many carnations would you need to produce 3 table centerpieces and 1 mantel decoration?

Suppose that carnations cost you 5 cents apiece and that you charge 3 cents labor for preparing each bloom. Cost and labor for each snapdragon stalk are given in Figure 5-4.

	cost	labor
carnations	5	3
snapdragons	6	4

FIGURE 5–4. EXPENSES OF BLOOMS.

What is the cost of the blooms for a table bouquet? The bouquet **EXAMPLE** uses 9 carnations at 5 cents apiece, or $9 \cdot 5$ cents, plus 4 snapdragons at 6 cents apiece, or $4 \cdot 6$ cents.

$$9 \cdot 5 + 4 \cdot 6 = 69 \text{ cents.}$$

Exercise 5–12. What is the cost of labor for the table bouquet?

Exercise 5–13. What is the cost of the blooms for a mantel bouquet?

Exercise 5–14. What is the cost of labor for a mantel bouquet?

Exercise 5–15. Figure 5–5 shows orders on four items from four businesses to a stationery supplier for

$$\begin{pmatrix} \text{placards} & \text{business cards} \\ \text{receipt pads} & \text{reorder pads} \end{pmatrix}$$

Zip Cafe	Zap Garage	Reilly Realty	Aunt Antiques
$\begin{pmatrix} 10 & 100 \\ 25 & 3 \end{pmatrix}$	$\begin{pmatrix} 4 & 100 \\ 10 & 4 \end{pmatrix}$	$\begin{pmatrix} 15 & 300 \\ 2 & 0 \end{pmatrix}$	$\begin{pmatrix} 6 & 500 \\ 5 & 0 \end{pmatrix}$

FIGURE 5–5. ORDERS TO STATIONERY SUPPLIER.

Which business ordered the most business cards? Which item was not ordered by two businesses? Which business ordered the most advertising placards?

Exercise 5–16. In Figure 5–5 find how many of each of the four items the stationery supplier will need to fill all four orders totaled.

MATRICES AND THEIR ARITHMETIC

The rectangular tables you have been using to record information are examples of 2 by 2 matrices. In general, r by s **matrices** result from tabulating information in a rectangle of r **rows** (horizontal lines) and s **columns** (vertical lines).*

EXAMPLES $\begin{pmatrix} 36'' & 24'' & 37'' \\ 38'' & 24'' & 36'' \end{pmatrix}$ A judge in a beauty contest has tabulated some notes on two contestants in the form of a matrix. The matrix has $r = 2$ rows, one for each girl, and $s = 3$ columns.

A disk jockey tabulates the playing times of tapes in a 20 by 1 matrix by listing the times in seconds beside the names on a list of the tapes.

A matrix with just one column is sometimes called a **column vector.**

$$\begin{array}{l|cc} \text{"Luv ya, Bob"} & 130 & \text{sec.} \\ \text{"Luv ya, Gal"} & 145 \\ \text{Soap-O Ad} & 100 \\ \text{Editorial} & 180 \\ \text{"Luv ya, Sue"} & 130 \\ \text{Charity Appeal} & 30 \\ \cdots\cdots\cdots & \cdots \\ \text{"Polka Blues"} & 135 \end{array}$$

This 1 by 4 matrix is an example of a **row vector,** that is, a matrix with just one row:

$$(5 \quad 10 \quad 15 \quad 20)$$

---

* The singular form is "matrix" (MAY-trix), plural "matrices" (MAY-tri-seez), adjective "matric" (MAY-trik).

Matrices have their own arithmetic, based on the information they represent. Two matrices are equal if their corresponding entries are equal. To correspond in all positions, two matrices have to have the same dimensions r and s.

$$\begin{pmatrix} 1.5 & 2 \\ -1 & 0 \end{pmatrix} = \begin{pmatrix} 3/2 & +\sqrt{4} \\ 9-10 & 0 \end{pmatrix}$$

$\begin{pmatrix} 1 & 2 & 3 \\ 4 & 5 & 6 \end{pmatrix}$ cannot equal $\begin{pmatrix} 1 & 2 \\ 4 & 5 \end{pmatrix}$ or any other 2 by 2 matrix, since

there are no corresponding entries to compare with 3 and 6.

Addition of matrices is based on totaling the categories of information tabulated in them, so the **sum** $A + B$ of two matrices is defined to be the matrix made up of sums of corresponding entries. Again, A and B must have the same dimensions r by s.

$$\begin{pmatrix} 2 & 1 \\ 3 & -1 \end{pmatrix} + \begin{pmatrix} -10 & 4 \\ 5 & 6 \end{pmatrix} = \begin{pmatrix} 2-10 & 1+4 \\ 3+5 & -1+6 \end{pmatrix} = \begin{pmatrix} -8 & 5 \\ 8 & 5 \end{pmatrix}.$$

Multiplication of matrices is based on combining related information as, for example, in Exercises 5–7, 5–8, 5–9, 5–12, 5–13, and 5–14. The **product** AB of matrices A and B has meaning when A breaks r groupings (such as the music organizations) into s subheadings (age groups) and B in turn breaks the s subheadings into t classifications (food portions). Then the product is an r by t matrix giving the r groupings in terms of the t classifications.

The matrix A breaks three factory items into two components each. **EXAMPLES**

	springs	bolts
thingum	5	2
whatsit	1	3
nubbin	0	1

$$A = \begin{pmatrix} 5 & 2 \\ 1 & 3 \\ 0 & 1 \end{pmatrix}.$$

The matrix B gives the costs of the components.

	cost
spring	1¢
bolt	$\frac{1}{2}$¢

$$B = \begin{pmatrix} 1 \\ \frac{1}{2} \end{pmatrix}.$$

Since A is 3 by 2 and B is 2 by 1, their product AB is 3 by 1.

$$AB = \begin{pmatrix} 5 \cdot 1¢ + 2 \cdot \frac{1}{2}¢ \\ 1 \cdot 1¢ + 3 \cdot \frac{1}{2}¢ \\ 0 \cdot 1¢ + 1 \cdot \frac{1}{2}¢ \end{pmatrix} = \begin{pmatrix} 6¢ \\ 2\frac{1}{2}¢ \\ \frac{1}{2}¢ \end{pmatrix}$$

gives the cost of components for each factory item.

Figure 5–6 shows a way to make matric multiplication so systematic that it can be turned over to a computer: copy the two tables with their entries carefully aligned. From two pieces of stiff paper, make guides, each with a slot that reveals just one line of a table at a time; the left guide reveals a horizontal line, or "row," the right guide a vertical line, or "column." As shown in Figure 5–6, the combination results are entered in a third "product" or "answer" table.

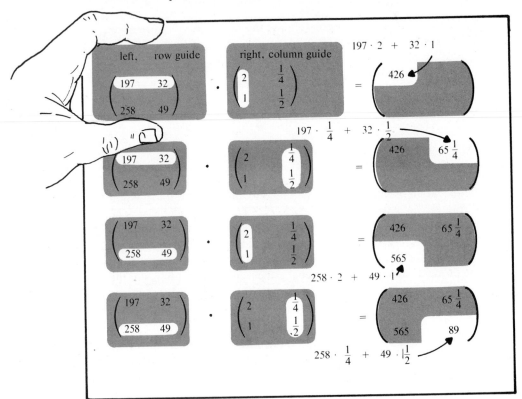

FIGURE 5–6. MULTIPLYING MATRICES USING A SLOTTED GUIDE.

Matric multiplication is simplified by slotted guides positioned to reveal one row in the left matrix and one column in the right matrix.

The numbers in the answer table are not labeled, but their positions in the table give their interpretations. For instance, the answer entry $65\frac{1}{4}$ comes from the first, or "band," row and the second, or "potato salad," column, so it represents the amount of potato salad planned for the band, both students and adults (Exercise 5–8).

In Figure 5–7 the "florist" tables of Figures 5–3 and 5–4 are combined.

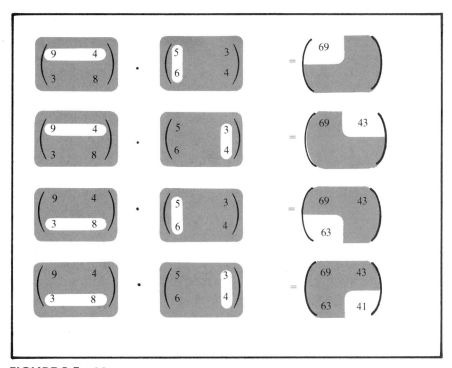

FIGURE 5–7. MATRIC MULTIPLICATION WITH SLOTTED GUIDES.

We use the same slotted guides used in Figure 5–6. The answer table gives the cost of blooms and the cost of labor for the table bouquet and for the mantel bouquet, summed in each case for whatever blooms (carnations or snapdragons) are used. For instance, the cost of labor for a mantel bouquet appears in the second "mantel" row, and the second "labor" column: $3 \cdot 3 + 8 \cdot 4 = 41 ¢$.

EXAMPLES

Let $A = \begin{pmatrix} 1 & 2 \\ 3 & 4 \end{pmatrix}$, $B = \begin{pmatrix} 5 & 6 \\ 7 & 8 \end{pmatrix}$, $C = (-1 \quad 3)$, $D = \begin{pmatrix} 2 \\ -1 \end{pmatrix}$, and

$$E = \begin{pmatrix} 1 & -1 & 1 \\ 5 & 1 & -1 \end{pmatrix}$$

Then $A + B = \begin{pmatrix} 6 & 8 \\ 10 & 12 \end{pmatrix}$.

$$CA = (-1 \cdot 1 + 3 \cdot 3 \quad -1 \cdot 2 + 3 \cdot 4) = (8 \quad 10).$$

Since A is 2 by 2 and C is 1 by 2, no product AC can be formed.

$$AE = \begin{pmatrix} 1(1) + 2(5) & 1(-1) + 2(1) & 1(1) + 2(-1) \\ 3(1) + 4(5) & 3(-1) + 4(1) & 3(1) + 4(-1) \end{pmatrix} = \begin{pmatrix} 11 & 1 & -1 \\ 23 & 1 & -1 \end{pmatrix}$$

$$BD = \begin{pmatrix} 5(2) + 6(-1) \\ 7(2) + 8(-1) \end{pmatrix} = \begin{pmatrix} 4 \\ 6 \end{pmatrix}.$$

Exercise 5–17. Find AD. Can DA be formed?

Exercise 5–18. Find AB and BA and compare them.

Matric multiplication is not commutative except in special cases. Even if A and B are square and of the same size so that AB and BA can both be formed, the products are usually different. This is because matrix multiplication is based on combining related information in a definite sequence.

Exercise 5–19. Let $A = \begin{pmatrix} 1 & 0 \\ -1 & \frac{1}{2} \end{pmatrix}$ and $B = \begin{pmatrix} 0 & 2 \\ 3 & \frac{1}{2} \end{pmatrix}$. Find $A + B$, $B + A$, AB, and BA.

Exercise 5–20. Let $A = \begin{pmatrix} 2 & 0 \\ 0 & 2 \end{pmatrix}$ and $B = \begin{pmatrix} 13 & 15 \\ 9 & 11 \end{pmatrix}$. Find AB and BA.

For Exercises 5–21 through 5–30, let $A = \begin{pmatrix} \frac{1}{2} & 5 \\ -1 & 2 \end{pmatrix}$, $B = \begin{pmatrix} 2 & 3 \\ 0 & 0 \end{pmatrix}$, $C = \begin{pmatrix} -2 & -3 \\ 0 & 1 \end{pmatrix}$, and $D = \begin{pmatrix} \frac{1}{2} & 0 \\ 0 & \frac{1}{2} \end{pmatrix}$.

Exercise 5–21. Find AB and BA.

Exercise 5–22. Find $A + B$.

Exercise 5–23. Find AC and CA.

Exercise 5–24. Find BC and CB.

Exercise 5–25. Find $C(A + B)$ and find CA and CB. If you add CA and CB, is the sum equal to $C(A + B)$?

Exercise 5–26. Find $(A + B)C$ and compare with $AC + BC$.

Exercise 5–27. Find AD and DA.

Exercise 5–28. Find BD and DB.

Exercise 5–29. Find CD and DC.

Exercise 5-30. Find $D^2 = DD$.

Exercise 5-31. Find $\begin{pmatrix} 1 & 4 \\ 1 & 3 \end{pmatrix} \begin{pmatrix} -3 & 4 \\ 1 & -1 \end{pmatrix}$ and $\begin{pmatrix} -3 & 4 \\ 1 & -1 \end{pmatrix} \begin{pmatrix} 1 & 4 \\ 1 & 3 \end{pmatrix}$.

Exercise 5-32. Find $\begin{pmatrix} -1 & 2 \\ 3 & -6 \end{pmatrix} \begin{pmatrix} 2 & 3 \\ 1 & \frac{3}{2} \end{pmatrix}$ and $\begin{pmatrix} 2 & 3 \\ 1 & \frac{3}{2} \end{pmatrix} \begin{pmatrix} -1 & 2 \\ 3 & -6 \end{pmatrix}$.

Exercise 5-33. Find $(1 \quad 2 \quad 3) \begin{pmatrix} 10 \\ 20 \\ 30 \end{pmatrix}$ and $\begin{pmatrix} 10 \\ 20 \\ 30 \end{pmatrix} (1 \quad 2 \quad 3)$.

Exercise 5-34. Demonstrate "associativity" of matric multiplication by finding

$$\left[\begin{pmatrix} 0 & 1 \\ 2 & 6 \end{pmatrix} \begin{pmatrix} 1 & 0 \\ 7 & 5 \end{pmatrix} \right] \begin{pmatrix} 2 \\ 3 \end{pmatrix} \quad \text{and} \quad \begin{pmatrix} 0 & 1 \\ 2 & 6 \end{pmatrix} \left[\begin{pmatrix} 1 & 0 \\ 7 & 5 \end{pmatrix} \begin{pmatrix} 2 \\ 3 \end{pmatrix} \right].$$

For the example in Figure 5–2, let A stand for the matrix $A = \begin{pmatrix} 197 & 32 \\ 258 & 49 \end{pmatrix}$ that tells how to get the music organization vector $\mathbf{m} = \begin{pmatrix} \text{band} \\ \text{glee} \end{pmatrix}$ from the age vector $\mathbf{v} = \begin{pmatrix} \text{students} \\ \text{adults} \end{pmatrix}$, by forming the matric product

$$\mathbf{m} = A\mathbf{v}.$$

Similarly, let B stand for the matrix $B = \begin{pmatrix} 2 & \frac{1}{4} \\ 1 & \frac{1}{2} \end{pmatrix}$ that tells how to get the

age vector $\mathbf{v}$ from the food vector $\mathbf{f} = \begin{pmatrix} \text{hot dogs} \\ \text{potato salad} \end{pmatrix}$, by forming the matric product

$$\mathbf{v} = B\mathbf{f}.$$

From $\mathbf{m} = A\mathbf{v}$ with the substitution $\mathbf{v} = B\mathbf{f}$, we see that

$$\mathbf{m} = A(B\mathbf{f}).$$

Matrices of numbers have an associative property under multiplication (illustrated in Exercise 5–34), which justifies regrouping the product $A(B\mathbf{f})$ so that the two matrices are taken together: $(AB)\mathbf{f}$. (Compare with the associative properties of a Boolean algebra in Definition 4–1.) Then the formula for $\mathbf{m}$ in terms of $\mathbf{f}$ becomes

$$\mathbf{m} = (AB)\mathbf{f}.$$

Exercise 5-35. Find AB and compare with Exercises 5–7, 5–8, and 5–9 to show that AB tells how to get $\mathbf{m}$ in terms of $\mathbf{f}$.

You can make the same sort of simplification for the information in Figures 5–3 and 5–4:

Exercise 5–36. Name the vector of floral arrangements (**f**, say, or **y**, or whatever appeals to you) and name the vector of types of blooms and the 2 by 2 matrix in Figure 5–3. Name the vector of costs and the 2 by 2 matrix in Figure 5–4. Show in matrix notation how to find the floral arrangements in terms of the costs. Compare with Exercises 5–12, 5–13, and 5–14.

You may have learned how to keep track of "dimensions" or "denominations" in applied problems by carrying them right along with your arithmetic, canceling or combining them as in this example:

EXAMPLE   How many seconds are there in 1.5 hours?

$$1.5 \text{ hrs.} = 1.5 \, \cancel{hrs.} \left(60 \, \frac{\text{min.}}{\cancel{hr.}} \right) \left(60 \, \frac{\text{sec.}}{\cancel{min.}} \right).$$

$$= 1.5(60)(60) \text{ sec., or } 5400 \text{ seconds.}$$

The vectors in the matric arithmetic

$$\mathbf{m} = A\mathbf{v}$$

$$\mathbf{v} = B\mathbf{f}$$

then $\mathbf{m} = A(B\mathbf{f}) = (AB)\mathbf{f}$

play a role like that of the denominations in ordinary multiplication problems. For instance, in Figure 5–2 the left matrix gives the music organizations in terms of the age grouping; the right matrix in turn gives the age groupings in terms of food portions. The product matrix then gives

$$\left(\frac{\text{music organizations}}{\text{age groups}} \right) \left(\frac{\text{age groups}}{\text{food portions}} \right) = \frac{\text{music organizations}}{\text{food portions}}$$

or music organizations in terms of food portions.

The matrices in Figure 5–8 summarize information about an ungraded primary school, where children 6, 7, and 8 years old are reading at three levels of proficiency, w.d. = with difficulty, ade. = adequately, and fl. = fluently. The left matrix L tells how many of each age **a** read at each level

		w.d.	ade.	fl.			bks.	wkbks.			cost	space
	6	48	23	7		w.d.	3	0		bks.	$2.47	$\frac{1}{2}''$
a age	7	8	50	22		p ade.	1	1		s wkbks.	0.70	$\frac{1}{4}''$
	8	8	45	31		fl.	4	2				
			a = Lp					p = Ms			s = Rc	

FIGURE 5–8. TABLED INFORMATION ON READING.

of proficiency **p**. The children who read with difficulty require 3 books and no workbook, the children who read adequately require 1 book and 1 workbook, and so on, as shown in the middle matrix M. The right matrix R gives the costs and the shelf space required for storage, **c**, of the supplies, **s**.

Exercise 5–37. Form the 3 by 2 product matrix MR. Read from it the answers to these questions: How much do supplies cost for a student who reads with difficulty? for one who reads fluently? Which supplies take up least shelf space: those for a poor reader, those for an adequate reader, or those for a fluent reader?

Exercise 5–38. Form the 3 by 2 product matrix LM. Which age group requires the most books? Which age group requires the most workbooks?

Exercise 5–39. Form the product matrix $(LM)R$ and the product matrix $L(MR)$ and note that they are equal. We have $\mathbf{a} = L\mathbf{p}$, $\mathbf{p} = M\mathbf{s}$, and $\mathbf{s} = R\mathbf{c}$, so by substituting,

$$\mathbf{a} = L\mathbf{p} = L(M\mathbf{s}) = LM(R\mathbf{c}) = LMR\mathbf{c}.$$

Thus the triple product $LMR = (LM)R = L(MR)$ gives age in terms of expenses. From the triple product matrix tell how much reading supplies cost for the 6-year-old students. How much do reading supplies cost for the 8-year-olds?

Exercise 5–40. Form the product AB of the matrices given in Figure 5–9. Interpret each entry of AB as information about the two Christmas tree farms.

	Trees			Economics	
	balsam firs	Scotch pines		average selling price	number of applications of herbicides
Farms farm 1	8000	10000	**Trees** b.f.	$1.75	10
farm 2	7000	6000	S.p.	$1.25	9
		A			B

FIGURE 5–9. INFORMATION ON TWO TREE FARMS.

Exercise 5–41. Form the product AB of the matrices given in Figure 5–10. Interpret each entry as information about baggage arrangements for a bus.

	Personnel athletes fans			**Baggage** carryons luggage compartment bags
Bus $\begin{matrix}1\\2\end{matrix}$	$\begin{pmatrix}12 & 18\\14 & 16\end{pmatrix}$		**Pers.** $\begin{matrix}\text{ath.}\\\text{fans}\end{matrix}$	$\begin{pmatrix}1 & \frac{1}{2}\\1\frac{1}{2} & 1\end{pmatrix}$
	A			B

FIGURE 5–10. BUS BAGGAGE.

Often a matrix of coefficients is fixed for a whole class of problems.

EXAMPLE A mail-order company offers these items:

	watch	coverall	bumper jack	wig	typing paper
price	$17.95	$6.37	$3.49	$28.95	$3.59
shipping weight	4 oz.	2 lb. 13 oz.	12 lb.	1 lb.	5 lb. 2 oz.

Give this 2 by 5 matrix of coefficients the name M, and notice that it stays fixed while the particular catalog or sale stays in effect. Now let the 5 by 1 column vector $\mathbf{v}$ list how many of each item, in order, a customer orders. For instance, John might order 1 watch, 2 coveralls, and 1 bumper jack, in which case his order-vector $\mathbf{v_J}$ would be

$$\mathbf{v_J} = \begin{pmatrix}1\\2\\1\\0\\0\end{pmatrix}.$$

The product $M\mathbf{v_J}$ of the 2 by 5 matrix and the 5 by 1 vector is the 2 by 1 matrix

$$M\mathbf{v_J} = \begin{pmatrix}\$34.18\\17\text{ lb. }14\text{ oz.}\end{pmatrix}.$$

The entries give the total cost of the items John orders and their shipping weight, respectively. Now let the column vector $\mathbf{v_M}$ tell how many of each item Margaret orders, say 1 watch, 1 wig, and 3 packages of typing paper.

Then $\mathbf{v_M} = \begin{pmatrix}1\\0\\0\\1\\3\end{pmatrix}$, and $M\mathbf{v_M} = \begin{pmatrix}\$57.67\\16\text{ lb. }10\text{ oz.}\end{pmatrix}$. The total cost for the

items Margaret orders is $57.67 and the shipping weight for her whole order is 16 lb. 10 oz.

Notice that the arithmetic is no different from totaling any order. We multiply each price or weight by the number of items ordered, then sum the products—a practical example of matrix multiplication.

You can imagine the listings in an entire mail-order catalog strung out in a 2 by several-thousand matrix, giving for each item the cost and the weight. Each order submitted would be a several-thousand by 1 column matrix or vector, composed mostly of 0's.

Exercise 5-42. Tickets to a production are $1.00 for adults, 55 cents for children. Let P be the 1 by 2 price matrix, the row vector

$$P = (\$1.00 \quad 0.55)$$

Let the column vector $\mathbf{v} = \begin{pmatrix} a \\ c \end{pmatrix}$ give the number of adults and children in a family, school class with adult supervisors, and so forth. Show that the product $P\mathbf{v}$ is a 1 by 1 matrix that gives for any vector $\mathbf{v}$ the total cost for tickets. Try it out on a group of 3 adults, 7 children; on a group of 12 adults, 1 child; on a group of 2 adults, 40 children. What happens if a is zero or if c is zero?

Air parcel post to Afghanistan from the United States costs $2.20 for **EXAMPLE** the first 4 oz. and $0.83 for each additional 4 oz. or fraction. Air parcel post to Great Britain from the United States costs $1.45 for the first 4 oz. and $0.36 for each additional 4 oz. or fraction.

$$\text{Let} \quad M = \begin{pmatrix} 2.20 & 0.83 \\ 1.45 & 0.36 \end{pmatrix}.$$

Let the weight vector $\mathbf{w}$ for a package be

$$\mathbf{w} = \begin{pmatrix} 1 \\ \text{number of additional 4 oz. or fraction after} \\ \text{the first 4 oz.} \end{pmatrix}$$

The product $M\mathbf{w}$, as the product of a 2 by 2 matrix and a 2 by 1 matrix, is a 2 by 1 matrix. Its entries give the cost of mailing a package with weight vector $\mathbf{w}$ by air parcel post from the United States to Afghanistan, and to Great Britain, respectively.

Exercise 5-43. Calculate the weight vector for a package weighing 2 pounds 2 ounces. Find the product matrix and interpret its two entries.

Exercise 5-44. Calculate the weight vector for a package weighing $3\frac{1}{2}$ oz. Find the product matrix and interpret its two entries.

Exercise 5-45. Calculate the weight vector for a package weighing $1\frac{1}{2}$ pounds. Find the product matrix and interpret its two entries.

Exercise 5–46. The matrix G in Figure 5–11 gives the grades of 4 students on each of 5 tests. The vector **w** gives the **weighting factors** the instructor wants to apply to the various tests, given in terms of the percentage of the course grade each represents (20% reserved for effect of homework, class participation, and so forth). Find the total (weighted) test grade for each student.

	1	2	3	4	5		
Abe	20	60	90	90	80		10%
Bob	70	90	80	100	80	$\mathbf{w} =$	15%
Cal	100	100	100	100	98		15%
Don	80	80	85	75	85		15%
							25%

FIGURE 5–11. WEIGHTING FACTORS.

Exercise 5–47. Let **v** be the 10 by 1 column vector

$$\begin{pmatrix} 1 \\ 2 \\ 3 \\ 4 \\ 5 \\ 6 \\ 7 \\ 8 \\ 9 \\ 10 \end{pmatrix}$$

and let **w** be the 1 by 10 row vector, $\mathbf{w} = (1\ 2\ 3\ 4\ 5\ 6\ 7\ 8\ 9\ 10)$. Then show that the matrix **vw** gives the body of the familiar multiplication table through 10 times 10.

Incidence Matrices

The matrix in Figure 5–12 is derived from the star; it has a 1 in the i, j position if point i and point j are connected by a line segment of the star, and a 0 otherwise. Even this simple a matrix contains a lot of information. Notice that its symmetry in rows and columns reflects the fact that in the star there is a line from i to j if and only if there is a line from j to i. Notice that there are exactly two 1's in each row and exactly two 1's in each column, since each point of the star is connected to exactly 2 other points.

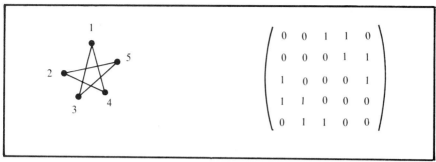

FIGURE 5–12. An incidence matrix.

The matrix has a 1 in row 3, column 5 to show that points 3 and 5 in the star are connected. The 2, 1 entry in the matrix is 0, because points 2 and 1 in the star are not connected.

The main diagonal of the matrix, that is, the i, i positions, 1, 1; 2, 2; and so forth, are all 0's, since no line segment connects a point to itself. The diagonal rows just above and just below the main diagonal are also composed entirely of 0's, because no star point is connected to the two adjacent points.

Perhaps you have already noticed that the 0's and 1's in the incidence matrix of Figure 5–12 can be interpreted either in their usual "counting" sense—"How many lines are there between i and j?"—or as simply "no" or "yes"—"Is there a line between i and j?"—or even as "false" or "true"—"A line segment connects points i and j: F or T." The "0" or "1" entries are especially convenient, though, because in some cases they can be summed or multiplied meaningfully.

Exercise 5–48. Arrange 6 points in a ring, spaced fairly evenly in a circular pattern. Number them in order. Then draw the lines indicated in the following incidence matrix.

	1	2	3	4	5	6
1	0	0	1	0	1	0
2	0	0	0	1	0	1
3	1	0	0	0	1	0
4	0	1	0	0	0	1
5	1	0	1	0	0	0
6	0	1	0	1	0	0

Exercise 5–49. Use an incidence matrix to describe the design in Figure 5–13. Comparison of Figure 5–13 with Figure 4–1 may suggest to you important applications of incidence matrices.

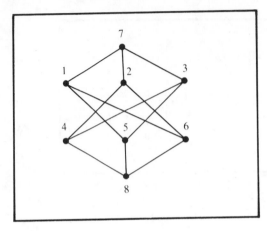

FIGURE 5-13. A BOOLEAN ALGEBRA DESIGN.

The next three exercises suggest the way incidence matrices can be used to describe a wide variety of situations.

Exercise 5-50. The following matrix has a "1" in position i, j if the ith girl in a club indicated on a questionnaire that she admires the jth girl.

$$
\begin{array}{cccc}
 & \text{Ada} & \text{Barb} & \text{Cass} & \text{Dot}
\end{array}
$$

	Ada	Barb	Cass	Dot
Ada	1	1	1	1
Barb	0	0	0	0
Cass	0	1	1	1
Dot	0	0	1	1

Which girl was most critical? least critical? Does Ada admire Dot? Does Dot admire Ada? What does the lack of symmetry in the matrix mean, interpreted in the social situation? How could you interpret the elements in the main diagonal positions 1, 1; 2, 2; 3, 3; and 4, 4?

Exercise 5-51. The banks in a small town have interlocking directorates, as shown in these lists of boards of directors.

Thirty-First National Bank: Appleby, Baum, Dinwiddie, Everett

Old Sock Bank: Baum, Cadwalleder, Dinwiddie, Fish

Cast Iron Bank: Appleby, Everett, Fish

River Bank: Appleby, Baum, Dinwiddie

Show in a 6 by 6 incidence matrix which people serve on some board together, and then extract this information from your matrix: What does the preponderance of "1's" indicate? Why do you expect the matrix to be symmetric? Which directors serve with every other director? Which director serves with the fewest others?

Exercise 5–52. A Public Health nurse has a chart showing for each clinic visitor which of four contagious diseases he has had: measles, mumps, chickenpox, and rubella. Invent such a chart for five of the visitors, in the form of an incidence matrix. Show how to find information from the matrix about the "incidence" of each disease.

As you can see, incidence matrices can be used very generally as check lists. Sometimes they can be combined meaningfully; that is, the product of two incidence matrices sometimes has a useful interpretation. The next two exercises are examples.

Exercise 5–53. Colleges *x*, *y*, and *z* offer various programs, as shown in matrix *A* of Figure 5–14. The programs, in turn, require various courses (among others), as shown in matrix *B*. Form the product *AB* and deduce which of the courses must be offered by each college.

	forestry	engineering	music			adv. calc.	botany	harmony
College *x*	1	1	1	for.		0	1	0
College *y*	1	0	1	engin.		1	0	0
College *z*	0	1	1	music		0	0	1
		A					*B*	

FIGURE 5–14. RELATED INCIDENCE MATRICES.

Exercise 5–54. Matrix *A* in Figure 5–15 shows individuals signed up for various programs. Matrix *B* shows various brochures and forms to be mailed to participants in the programs. Write the product matrix *AB* and interpret its entries.

	swim	hike	boat	nature study			hlth. form	swim O.K.	backpack list	collector's list
Joe	1	0	1	0		swim	1	0	0	0
Moe	0	1	0	1		hike	0	0	1	0
Dan	1	1	1	0		boat	0	1	0	0
			A			nature study	0	0	0	1
								B		

FIGURE 5–15. TWO RELATED INCIDENCE MATRICES.

Incidence matrices can be used as truth tables. Exercise 5–55 gives you a special case to have in mind during the development of this idea.

Exercise 5–55. In Figure 5–16 six points are arranged on 4 lines. The points are numbered 1 through 6, the lines lettered d, e, f, g. Show in a 4 by 6 incidence matrix which lines lie on which points, or equivalently which points lie on which lines.

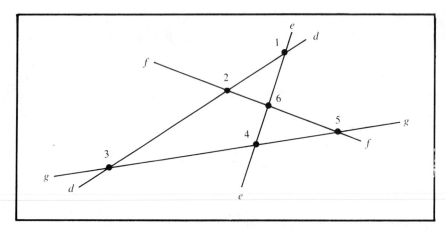

FIGURE 5–16. A SIX-POINT GEOMETRY.

In Exercise 5–55 you formed a 4 by 6 matrix with each of the four rows standing for one of the lines in Figure 5–16. Another way to interpret the same matrix would be this: Let the first row stand for the logical statement "x is on the line d." As in Chapter 3, a "logical statement" means simply a statement that must be either True or False but not both for each x. Let the second row stand for "x is on the line e," and so on. Let the columns stand for the 6 points and place a "1" in the i, j position if the ith statement is True when x is replaced by j. Place a "0" in the i, j position if the ith statement has the truth value "False" when x is replaced by j.

In general, let $P_1(x), P_2(x), \ldots, P_i(x), \ldots$ be logical statements about x, where x can take on values in a certain universe of discourse, or space, U. Let each row in a matrix stand for one of the statements and let each column stand for one of the elements of the space U. An incidence matrix i, j can reflect the truth tables for all the listed statements at once if it has a "1" in the i, j position if $P_i(j)$ is True, a "0" if $P_i(j)$ is False.

EXAMPLE   Let $P(x)$ be the logical statement: "x has a scholarship." Let $Q(x)$ be: "x has a job." $R(x)$: "x is independently wealthy." $S(x)$: "x takes a full load of courses." $T(x)$: "x takes a partial load of courses." Let the space U be the students classified as freshmen a certain semester at a certain college. Then we can show all the relevant information compactly in a matrix like that of Figure 5–17.

			U				
	Dan	Beth	Barb.	Bob	Nate	Son	Cher
$P(x)$	1	0	0	0	0	1	0
$Q(x)$	1	0	1	1	1	1	0
$R(x)$	0	0	0	0	1	0	1
$S(x)$	1	1	1	0	0	1	1
$T(x)$	0	0	0	1	1	0	0

FIGURE 5–17. STUDENT FINANCES.

Exercise 5–56. From Figure 5–17, which students are independently wealthy? Do any of these wealthy students hold scholarships? Do any of them have jobs? Do any scholarship holders take only a partial load of courses? From the interpretations of $S(x)$ and of $T(x)$, what would you say about the logical statement $S(x) \vee T(x)$, that is, "$S(x)$ or $T(x)$ or both" for x in U? Show how Figure 5–17 bears this out. What would you say about $S(x) \wedge T(x)$? Does the incidence matrix show this feature? Express the information about scholarship holders versus wealthy students by connecting $P(x)$ and $R(x)$ as logical statements.

In Chapter 4 spaces of logical statements were linked to spaces of sets. Figure 5–17 can be reinterpreted in terms of sets very easily. Let P be the set of students in the space U who hold scholarships, Q be the set of students in U who have jobs, and so forth. Then there is a 1 in the i, j position if the jth student is a member of the ith set. For further practice try interpreting your matrix of Exercise 5–55 in terms of sets.

AN INVERSE MATRIX

You are a planner for the Steel-O Construction Company, which markets houses made of prefabricated modular room units. You have two house designs, the Pad, which uses two bedroom modules and one bathroom module, and the Gables, which uses three bedroom modules and two bathroom modules.

Equation (1) presents this information in compact matrix form

$$\mathbf{h} = A\mathbf{m}, \tag{1}$$

by letting $\mathbf{h}$ represent a "house" vector and $\mathbf{m}$ a "module" vector,

$$\mathbf{h} = \begin{pmatrix} \text{Pad} \\ \text{Gables} \end{pmatrix}; \quad \mathbf{m} = \begin{pmatrix} \text{bedrooms} \\ \text{bathrooms} \end{pmatrix}; \quad \text{and } A \text{ the matrix}$$

$$A = \begin{pmatrix} 2 & 1 \\ 3 & 2 \end{pmatrix}.$$

Then the matric equation (1) summarizes the information

$$\begin{pmatrix} \text{Pad} \\ \text{Gables} \end{pmatrix} = \begin{pmatrix} 2 & 1 \\ 3 & 2 \end{pmatrix} \begin{pmatrix} \text{bedrooms} \\ \text{bathrooms} \end{pmatrix}.$$

Suppose for a moment that formula (1) involved ordinary multiplication of numbers, not matrices. For instance, the formula $h = 4m$ might give the number of lighting fixtures h in terms of the number of modules m. In this case, to find m in terms of h you could find an inverse $(4)^{-1} = \frac{1}{4}$ for 4, with the property that $(\frac{1}{4})4 = 1$, so that, multiplying $h = 4m$ by $\frac{1}{4}$, you could find

$$\tfrac{1}{4}h = (\tfrac{1}{4})4m = m.$$

An inverse A^{-1} for the matrix A analogous to the inverse $\frac{1}{4}$ for 4 would make it possible to find the vector $\mathbf{m}$ in terms of the vector $\mathbf{h}$. From the point of view of your work as planner for Steel-O Construction, equation (1), $\mathbf{h} = A\mathbf{m}$, tells you how many modules of each type will be needed for any planned number of Pads and Gables. An equation $\mathbf{m} = A^{-1}\mathbf{h}$ would tell you what changes to make in Pad and Gable planning to use up a supply of prefabricated modules of the two types.

Since the inverse matrix to A is a matrix that multiplies A to give the **identity matrix**, the first step is to find what matrix plays the role of "1" or multiplicative identity for matrices. The 2 by 2 identity matrix is

$$I_2 = \begin{pmatrix} 1 & 0 \\ 0 & 1 \end{pmatrix}.$$

Exercise 5–57. Verify that

$$I_2 \begin{pmatrix} 1 & 3 \\ 2 & -4 \end{pmatrix} = \begin{pmatrix} 1 & 0 \\ 0 & 1 \end{pmatrix} \begin{pmatrix} 1 & 3 \\ 2 & -4 \end{pmatrix} = \begin{pmatrix} 1 & 3 \\ 2 & -4 \end{pmatrix}.$$

Exercise 5–58. Verify that $\begin{pmatrix} 2 & -2 \\ 1 & 3 \end{pmatrix} I_2 = \begin{pmatrix} 2 & -2 \\ 1 & 3 \end{pmatrix}.$

Exercise 5–59. Let $\begin{pmatrix} a & b \\ c & d \end{pmatrix}$ stand for a 2 by 2 matrix A. Then $I_2 A$ has first row $\begin{pmatrix} 1 \cdot a + 0 \cdot c & 1 \cdot b + 0 \cdot d \end{pmatrix}$, or $\begin{pmatrix} a & b \end{pmatrix}$. Find the second row of $I_2 A$ and find $A I_2$.

More generally, the **r by r identity matrix** I_r has a one as each main diagonal entry, and all zeros elsewhere:

$$I_r = \begin{pmatrix} 1 & 0 & 0 & \cdots & 0 \\ 0 & 1 & 0 & \cdots & 0 \\ 0 & 0 & 1 & \cdots & 0 \\ & & \cdots & & \\ 0 & 0 & 0 & \cdots & 1 \end{pmatrix}.$$

Let $A = \begin{pmatrix} a & b \\ c & d \end{pmatrix}$ represent any 2 by 2 matrix. Calculate the quantity $ad - bc$. This quantity is called the **determinant** of the matrix A and is often designated by Δ or by $|A| = \begin{vmatrix} a & b \\ c & d \end{vmatrix}$.

$$\Delta = |A| = ad - bc.$$

If the determinant is not zero, then A has an **inverse matrix,** that is, a matrix A^{-1} such that $A^{-1}A = AA^{-1} = I_2$. If the determinant of A is zero, then there is no matrix that can multiply A to give the identity matrix.

Suppose the determinant Δ is not zero, so that A has an inverse matrix. Then the inverse can be found to be

$$A^{-1} = \begin{pmatrix} \dfrac{d}{\Delta} & \dfrac{-b}{\Delta} \\ \dfrac{-c}{\Delta} & \dfrac{a}{\Delta} \end{pmatrix}.$$

By actual calculation you can verify that with A^{-1} defined in this way, $AA^{-1} = I_2$, and also $A^{-1}A = I_2$.

You may wonder where the form for the inverse comes from. Suppose the matrix A summarizes information such as

$$\begin{cases} \text{Pad} = a \text{ bedrooms} + b \text{ bathrooms} \\ \text{Gables} = c \text{ bedrooms} + d \text{ bathrooms} \end{cases}$$

Multiply the Pad equation by d, the Gables equation by $-b$, and add, getting $(ad - bc)$ bedrooms $= d$ Pads $+ (-b)$ Gables, so that

$$1 \text{ bedroom} = \left(\dfrac{d}{\Delta}\right) \text{Pads} + \left(\dfrac{-b}{\Delta}\right) \text{Gables.}$$

Similarly, multiply the Pad equation by $-c$, the Gables equation by a, and add to find

$$1 \text{ bathroom} = \left(\dfrac{-c}{\Delta}\right) \text{Pads} + \left(\dfrac{a}{\Delta}\right) \text{Gables.}$$

Let $A = \begin{pmatrix} 1 & 2 \\ -3 & 4 \end{pmatrix}$. Then $\Delta = ad - bc = 1 \cdot 4 - 2(-3) = 10 \neq 0$, **EXAMPLE** so form

$$A^{-1} = \begin{pmatrix} \dfrac{4}{10} & \dfrac{-2}{10} \\ \dfrac{3}{10} & \dfrac{1}{10} \end{pmatrix}.$$

Then $AA^{-1} = \begin{pmatrix} 1 & 2 \\ -3 & 4 \end{pmatrix}\begin{pmatrix} .4 & -.2 \\ .3 & .1 \end{pmatrix}$

$$= \begin{pmatrix} 1(.4) + 2(.3) & 1(-.2) + 2(.1) \\ -3(.4) + 4(.3) & -3(-.2) + 4(.1) \end{pmatrix} = \begin{pmatrix} 1 & 0 \\ 0 & 1 \end{pmatrix} = I_2.$$

Exercise 5–60. Show that $\begin{pmatrix} .4 & -.2 \\ .3 & .1 \end{pmatrix}\begin{pmatrix} 1 & 2 \\ -3 & 4 \end{pmatrix} = I_2$.

Exercise 5–61. Let A be $\begin{pmatrix} 3 & 4 \\ 6 & -2 \end{pmatrix}$. Form A^{-1} and multiply it by A as a check. The product should be I_2.

Exercise 5–62. Let $A = \begin{pmatrix} 1 & 2 \\ 3 & 4 \end{pmatrix}$. Find A^{-1}.

Exercise 5–63. Let $B = \begin{pmatrix} 2 & 1 \\ 0 & 2 \end{pmatrix}$. Find B^{-1}.

Exercise 5–64. Let $C = \begin{pmatrix} 2 & 1 \\ -1 & 6 \end{pmatrix}$. Find C^{-1}.

Exercise 5–65. Let $D = \begin{pmatrix} 0 & 1 \\ 1 & 0 \end{pmatrix}$. Find D^{-1}.

Exercise 5–66. Let $S = \begin{pmatrix} s & 0 \\ 0 & s \end{pmatrix}$, where s stands for some non-zero number. Find S^{-1}.

Exercise 5–67. Let $E = \begin{pmatrix} 3.1 & 1.4 \\ 2.3 & 1 \end{pmatrix}$. Find E^{-1}, carrying 2 decimal places (that is, rounding calculations to the nearest hundredth). Check the product $E^{-1}E$. How accurate are its entries as approximations to the entries 1 and 0 of I_2?

Returning to formula (1), $\mathbf{h} = A\mathbf{m}$, with $A = \begin{pmatrix} 2 & 1 \\ 3 & 2 \end{pmatrix}$, you can now find $A^{-1} = \begin{pmatrix} 2 & -1 \\ -3 & 2 \end{pmatrix}$. Multiplying both sides of (1) by A^{-1}:

$$A^{-1}\mathbf{h} = A^{-1}(A\mathbf{m}) = (A^{-1}A)\mathbf{m} = I_2\mathbf{m} = \mathbf{m},$$

or

$$\begin{pmatrix} 2 & -1 \\ -3 & 2 \end{pmatrix}\begin{pmatrix} \text{Pad} \\ \text{Gables} \end{pmatrix} = \begin{pmatrix} \text{bedrooms} \\ \text{bathrooms} \end{pmatrix}.$$

From this matric equation,

$$\begin{cases} 1 \text{ bedroom} = 2 \text{ Pads} - 1 \text{ Gables} \\ 1 \text{ bathroom} = -3 \text{ Pads} + 2 \text{ Gables} \end{cases}$$

The first formula means that an increase of 2 Pad houses and a decrease of one Gables house will result in use of one more bedroom module with no change in the number of bathrooms needed. According to the second formula, a decrease of 3 Pads with an increase of 2 Gables will call for one more bathroom, with no change in the bedroom module requirements.

Exercise 5–68. Each unit of boating requires 8 hours and costs \$8. Each unit of singing requires 3 hours and costs \$2. Write this as

$$\mathbf{f} = B\mathbf{r},$$

$$\begin{pmatrix} \text{boating} \\ \text{singing} \end{pmatrix} = \begin{pmatrix} 8 & 8 \\ 3 & 2 \end{pmatrix} \begin{pmatrix} \text{hours} \\ \text{dollars} \end{pmatrix}.$$

Find B^{-1} and form $B^{-1}\mathbf{f} = B^{-1}B\mathbf{r} = I_2\mathbf{r} = \mathbf{r}$. What change in your activity schedule, boating and singing, will result in use of one more hour with no change in cost? Then suppose you need to spend 4 *fewer* hours on recreation without changing the cost. What adjustment would make that possible?

Exercise 5–69. Suppose that you are a toy manufacturer. In Figure 5–18 each Toy Town has one Long Building and three Square Buildings. Each Toy Village has two Long Buildings and one Square Building. Each Long Building requires 12 cube blocks and 2 triangle blocks. Each Square Building requires 8 cube blocks and 2 triangle blocks. Express this information in two matrices. Multiply the matrices and interpret the product.

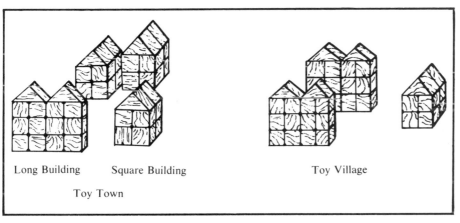

Long Building   Square Building   Toy Village

Toy Town

FIGURE 5–18.

Exercise 5–70. In the preceding exercise, verify that the inverse of the matrix that gives the block requirements for each building is

$$\begin{pmatrix} \frac{1}{4} & -\frac{1}{4} \\ -1 & \frac{3}{2} \end{pmatrix}.$$

Then to use exactly 4 more cube blocks with no change in the triangle requirements, build 1 more Long Building and __ fewer Square Buildings. To use exactly 2 more triangle blocks with no change in cube requirements, build __ fewer Long Buildings and __ more Square Buildings.

An introduction to matrices and determinants appears in Chapter 8 of *The Nature of Mathematics* by Frederick H. Young (John Wiley & Sons, Inc., 1968).

REVIEW OF CHAPTER 5

1. Find $A + B$, AB, BA, $|A|$, A^{-1}, $|B|$, and B^{-1}, for the 2 by 2 matrices

$$A = \begin{pmatrix} 1 & 2 \\ 0 & 1 \end{pmatrix} \quad \text{and} \quad B = \begin{pmatrix} 3 & 2 \\ -1 & 1 \end{pmatrix}.$$

2. Your job involves a study of low-income housing, specifically some apartment houses containing one 3-bedroom apartment and six 1-bedroom apartments each, and duplexes containing one 3-bedroom apartment and one 1-bedroom apartment each. Express the building vector

$$\mathbf{b} = \begin{pmatrix} \text{apartment house} \\ \text{duplex} \end{pmatrix}$$

as a housing matrix H times the size vector

$$\mathbf{s} = \begin{pmatrix} \text{3-bdr.} \\ \text{1-bdr.} \end{pmatrix}$$

3. Suppose that on the average a 3-bedroom apartment provides housing for 2 adults and 5 children, and that a 1-bedroom apartment provides housing for 2 adults and 2 children. Summarize this information using a personnel matrix P and an age vector

$$\mathbf{a} = \begin{pmatrix} \text{adults} \\ \text{children} \end{pmatrix}.$$

4. Find $\mathbf{b}$ in terms of $\mathbf{a}$, calculating the necessary product matrix $R = HP$, and interpreting each of its entries.

5. Find R^{-1} and use it to express $\mathbf{a}$ in terms of $\mathbf{b}$.

6. Suppose that you need to change the proposed housing plans so as to shelter 15 more children without changing the provision for adults. Use the second row of R^{-1} to find how many fewer apartment houses and how many more duplexes you should plan for.

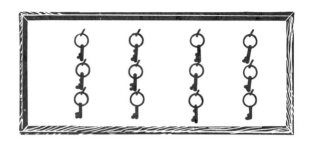

CHAPTER 5 AS A KEY TO MATHEMATICS

Even in this short introduction you can recognize the surprising power of methodical tabulation and system in solving problems. In Chapter 10 we find that system and format are basic to use of computers, too.

Matrices provide us with another example of an arithmetic to compare and contrast with others, such as familiar real-number arithmetic, modular arithmetic (Chapter 2), and Boolean algebra (Chapter 4).

Perhaps nowhere is the steady-by-jerks history of mathematical ideas better illustrated than in the development of matrices. Just as a tool-and-die maker makes tools for toolmakers, and an education teacher teaches teachers to teach, so mathematicians develop many subjects for "internal" use, that is, for use in developing more mathematics. The English mathematician Arthur Cayley (1821–1895) introduced matrices in 1858, as an "internal" mathematical contribution. Then Heisenberg in 1925, 67 years after Cayley's memoir, found matrices all polished and waiting, as if tailored for use in quantum mechanics. The determinant of a matrix has had a separate history, only comparatively recently seen in its modern relation to matrices.

CHAPTER 6

But Perseus went on boldly, past many an ugly sight, far away into the heart of the Unshapen Land, beyond the streams of Ocean, to the isles where no ship cruises, where is neither night nor day, where nothing is in its right place, and nothing has a name.

Heroes: Greek Fairy Tales, *edited by Charles Kingsley* *

Chapters 2, 4, and 5 give examples of arithmetics, arithmetics which have various features and are appropriate for attacking various problems. In this chapter we see how structures are discerned in mathematics, how they are named, illustrated, compared, and contrasted. Here we tie together two of the most important aspects of mathematics: the abstract character of mathematical models, and the function of mathematics as a language.

* Schocken Books, Inc., 1970.

GROUPS, RINGS, FIELDS

Classifications of Some Mathematical Systems

GROUPS

You and two coworkers, Max and Zelda, have found a way to relieve the monotony of a repetitious part of your work, by exchanging jobs at intervals. For instance, in one exchange, which can be called a, you (Y) take over the job Max (X) has been doing, Zelda (Z) takes over your previous job, and Max takes over the job Zelda has been doing. This information can be condensed to the short form:

$$a = \begin{pmatrix} X & Y & Z \\ Y & Z & X \end{pmatrix}.$$

The vertical columns can be written in any order; for instance, a could be written

$$a = \begin{pmatrix} Z & X & Y \\ X & Y & Z \end{pmatrix} \text{ or } a = \begin{pmatrix} Y & X & Z \\ Z & Y & X \end{pmatrix}, \text{ and so on.}$$

In exchange

EXAMPLE

$$b = \begin{pmatrix} X & Y & Z \\ X & Z & Y \end{pmatrix},$$

Max keeps whatever job he has been doing, and you and Zelda trade jobs, for from $\begin{pmatrix} - & Y & - \\ - & Z & - \end{pmatrix}$, Zelda takes over your job, and from $\begin{pmatrix} - & - & Z \\ - & - & Y \end{pmatrix}$, you take over Zelda's. Since X appears directly under X, Max replaces himself.

155

Exercise 6–1. Interpret

$$c = \begin{pmatrix} X & Y & Z \\ Y & X & Z \end{pmatrix}.$$

Exercise 6–2. Interpret

$$d = \begin{pmatrix} X & Y & Z \\ Z & X & Y \end{pmatrix}.$$

Exercise 6–3. Fill in the second row of

$$e = \begin{pmatrix} X & Y & Z \\ & & \end{pmatrix}$$

so as to form the "trivial" exchange, in which each person keeps his same job; that is, X replaces X, Y replaces Y, and Z replaces Z.

Exercise 6–4. Complete the second row of

$$f = \begin{pmatrix} X & Y & Z \\ & & \end{pmatrix}$$

so that you (Y) keep the same job, but Max and Zelda trade jobs.

The 6 exchanges a, b, c, d, e, f are all the possible ways the 3 coworkers can exchange the 3 jobs. Notice that you do not need to know which job each worker started with; in fact, you do not even have a description of the three jobs.

EXAMPLE

Exchange $f = \begin{pmatrix} X & Y & Z \\ Z & Y & X \end{pmatrix} = \begin{pmatrix} Y & X & Z \\ Y & Z & X \end{pmatrix} = \begin{pmatrix} Z & X & Y \\ X & Z & Y \end{pmatrix}$, and so

forth, can be applied meaningfully to various situations:

$$f \begin{pmatrix} X \text{ filing} \\ Y \text{ typing} \\ Z \text{ telephoning} \end{pmatrix} \rightarrow \begin{pmatrix} Z \text{ filing} \\ Y \text{ typing} \\ X \text{ telephoning} \end{pmatrix}$$

$$f \begin{pmatrix} Y \text{ telephoning} \\ X \text{ typing} \\ Z \text{ filing} \end{pmatrix} \rightarrow \begin{pmatrix} Y \text{ telephoning} \\ Z \text{ typing} \\ X \text{ filing} \end{pmatrix}$$

$$f \begin{pmatrix} Z \text{ shaking hands} \\ X \text{ kissing babies} \\ Y \text{ giving speeches} \end{pmatrix} \rightarrow \begin{pmatrix} X \text{ shaking hands} \\ Z \text{ kissing babies} \\ Y \text{ giving speeches} \end{pmatrix}$$

The **composition** $g \circ h$ of two exchanges is the exchange equivalent to applying h and then g.

EXAMPLE

$$a \circ f = \begin{pmatrix} X & Y & Z \\ & \downarrow & \\ Y & Z & X \end{pmatrix} \circ \begin{pmatrix} X & Y & Z \\ & \downarrow & \\ Z & Y & X \end{pmatrix} = \begin{pmatrix} X & Y & Z \\ X & Z & Y \end{pmatrix} = b,$$

because in exchange f the job Max started with goes to Zelda, then in exchange a back to Max; in exchange f Your job goes to You, then in exchange a to Zelda; in exchange f Zelda's job goes to Max, then in exchange a from Max to You. Then the net effect of the two exchanges in sequence is

$$\begin{pmatrix} X & Y & Z \\ X & Z & Y \end{pmatrix},$$

which is the exchange labeled b. The first row of b is taken in the order X, Y, Z, for convenient reference, but the order is arbitrary.

EXAMPLE

$$b \circ a = \begin{pmatrix} X & Y & Z \\ X & Z & Y \end{pmatrix} \circ \begin{pmatrix} X & Y & Z \\ Y & Z & X \end{pmatrix} = \begin{pmatrix} X & Y & Z \\ Z & Y & X \end{pmatrix} = f.$$

To find the net effect on each letter, follow it through the right exchange a, then follow the result through the left exchange b:

$$\begin{pmatrix} - & Y & - \\ & & \\ - & Z & - \end{pmatrix} \circ \begin{pmatrix} X & - & - \\ & \downarrow & \\ Y & - & - \end{pmatrix} = \begin{pmatrix} X & - & - \\ & & \\ Z & - & - \end{pmatrix}$$

$$\begin{pmatrix} - & - & Z \\ & & \\ - & - & Y \end{pmatrix} \circ \begin{pmatrix} - & Y & - \\ & & \\ - & Z & - \end{pmatrix} = \begin{pmatrix} - & Y & - \\ & & \\ - & Y & - \end{pmatrix}$$

$$\begin{pmatrix} X & - & - \\ & & \\ X & - & - \end{pmatrix} \circ \begin{pmatrix} - & - & Z \\ & & \\ - & - & X \end{pmatrix} = \begin{pmatrix} - & - & Z \\ & & \\ - & - & X \end{pmatrix}$$

Exercise 6–5. Show that $a \circ b = c$: First find the net effect of $a \circ b$ on X. The right exchange b replaces X by X; then the left exchange a replaces X by Y. Then

$$a \circ b = \begin{pmatrix} X & Y & Z \\ Y & & \end{pmatrix},$$

with two positions yet to be determined. Now find the net effect of $a \circ b$ on Y. Exchange b replaces Y by Z; then exchange a replaces that result, Z, by X. Check to be sure that the net effect of $a \circ b$ on Z is to replace it by Z.

FIGURE 6-1. COMPOSITION $b \circ f = a$.

Three students exchange jobs symbolized by a camera, a tennis racket, and a stethoscope. Under exchange f, X, and Z exchange jobs. Then under exchange b, Y and Z exchange jobs. The net effect of the composition is the same as exchange a, which gives X's job to Y, Y's job to Z, and Z's job to X.

Then write

$$a \circ b = \begin{pmatrix} X & Y & Z \\ Y & X & Z \end{pmatrix},$$

and identify this as the exchange called c.

Exercise 6–6. Find $f \circ a$.

Exercise 6–7. Find $d \circ f$.

Exercise 6–8. Find $f \circ d$.

Exercise 6–9. Show that $a \circ a = d$.

Exercise 6–10. Show that $a \circ (a \circ a) = a \circ d = e$.

Exercise 6–11. Show that $d \circ a = e$.

Exercise 6–12. Show that $b \circ b = e$.

Exercise 6–13. Show that $c \circ c = e$.

Exercise 6–14. Show that $f \circ f = e$.

Exercise 6–15. Explain the results of Exercises 6–12, 6–13, and 6–14 in terms of job exchanges.

Exercise 6–16. Show that whatever exchange g is (that is, one of a, b, c, d, e, or f), $g \circ e = g$ and $e \circ g = g$.

The compositions you have been finding can be recorded compactly in a **Cayley* operation table,** which is much like a multiplication table, except that it can record operations other than multiplication, in this case composition (Figure 6–2). The first row and column are labeled e for the **identity** exchange that does not change any of the 6 exchanges. (See Exercise 6–16.) In the horizontal row labeled g and the vertical column labeled h, the entry is the composition $g \circ h$; that is, the net result of h followed by g. For instance, from Exercise 6–5 the entry in the a row under the column heading b should be c.

Exercise 6–17. Enter in the Cayley table of Figure 6–2 the results of Exercises 6–6, 6–7, and 6–8. Find $a \circ c$ and $c \circ a$ and enter them.

---

* This is the same Arthur Cayley who introduced matrices in 1858. (See page 153).

∘	e	a	b	c	d	f
e	e	a	b	c	d	f
a	a	d	c		e	b
b	b	f	e			
c	c			e		
d	d	e				
f	f					e

FIGURE 6–2. COMPOSITION TABLE FOR SIX JOB EXCHANGES.

Exercise 6–18. In Figure 6–2 complete the b row and the b column.

Exercise 6–19. Find $d \circ c$, and enter the result in Figure 6–2.

Exercise 6–20. Find $d \circ d$.

Exercise 6–21. Find $f \circ c$.

Exercise 6–22. Complete the operation table of Figure 6–2 and as a check make sure that each exchange appears exactly once in each row and exactly once in each column.

Symmetries of a Triangle

The arithmetic developed for 6 exchanges of jobs among 3 workers also describes the rotations and flips of an equilateral triangle illustrated in Figure 6–3. Exchange b stands for

$$b = \begin{pmatrix} X & Y & Z \\ X & Z & Y \end{pmatrix},$$

so for the triangle it stands for a flip of the triangle around its bisector through X, that is, a flip with vertex X held fixed. The identity e leaves the triangle in whatever orientation it has initially. Exchanges a and d are represented by rotations of the triangle. Exchange c is a flip with vertex Z as pivot; f is a flip with Y as pivot.

Make a paper triangle with one side shaded and label the vertices as in Figure 6–3. The 6 ways the triangle can be lifted and replaced to cover the same space as before are called "symmetries" of the triangle. Now use the triangle as a computing device to form the composition $a \circ f$: First, flip the triangle with Y as pivot, to represent f, interchanging vertices X and Z. Then rotate the triangle so that vertex X is replaced by Y, Y by Z, and Z by X, as called for in a. Show that the net change in the triangle could be brought about equivalently by b, a flip with vertex X held fixed.

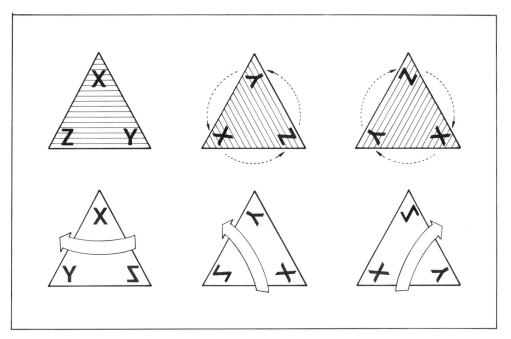

FIGURE 6–3. THE SIX SYMMETRIES OF A TRIANGLE.

There are six different "symmetries" of an equilateral triangle (ways it can be lifted and replaced to cover the same area). The identity symmetry replaces the triangle with no change. Two symmetries rotate the triangle and three flip it over.

Exercise 6–23. Use the paper triangle to demonstrate $a \circ a = d$, and $d \circ d = a$.

Exercise 6–24. Describe in terms of the triangle why $a \circ a \circ a = e$.

Exercise 6–25. Reinterpret Exercise 6–15 for the triangle.

Exercise 6–26. Find $f \circ a$ and show that it is different from $a \circ f$.

The 6-element mathematical system you have been studying is an example of a *group*, and its combining rule, composition, is an example of a *binary operation*. Since a group involves not just a set of elements, but also a binary operation for combining them, we define such an operation before we define a group:

Definition 6–1. A **binary operation** $*$ on a mathematical system of elements a, b, c, . . . is a combining rule that gives for each two elements a, b exactly one answer element $a * b$ in the system. ("Binary" means "two-ary" and refers to the fact that the operation combines elements two at a time.)

EXAMPLES   The combining rule + (addition) is a binary operation on the integers, because for every two integers i, j there is exactly one sum, or answer, element $i + j$, and that element is again an integer:

$$2 + (-1) = 1, \quad (-7) + 12 = 5, \quad 10 + 100 = 110, \quad 10 + (-10) = 0,$$

and so on.

The combining rule × (multiplication) is also a binary operation on the integers, because for every two integers i, j there is exactly one product, or answer, $i \times j$, and that answer is again an integer:

$$2 \times (-1) = -2, \quad (-7) \times 12 = -84, \quad 10 \times 100 = 1000,$$
$$10 \times 0 = 0, \text{ and so on.}$$

Human "multiplication" or any bisexual reproduction is not a binary operation, for two reasons: In the first place, no "answer" or offspring can be assigned to some pairs, such as male * male or Eve * Eve. In the second place, the offspring are not members of the original "parent" set.

Take a set of seven swimmers and let each compete with each of the others once, the "answer" for each match being the winner of the match. Suppose the answer in case a swimmer s competes with himself is by convention called s. Here we have a binary operation, because every pair $s * t$ yields exactly one winner and that winner is one of the seven swimmers.

Define a combining rule ⓨ on the people now in this building, as follows:

$$a \text{ⓨ} b = \text{the one of } a \text{ or } b \text{ that is the older } in \text{ } years.$$

For instance,

$$\text{instructor} \text{ⓨ} \text{student} = \text{instructor}$$

$$\text{senior} \text{ⓨ} \text{freshman} = \text{senior, in most cases}$$

For most building populations, the rule ⓨ fails to be a binary operation, because the answer to a question $a \text{ⓨ} b = ?$ can be ambiguous. Since the ages are measured in years, there will probably be pairs a, b of people having the same age. In such a case, $a \text{ⓨ} b$ could be either a or b or both. Then ⓨ is not a binary operation. Notice that a failure in *any* instance keeps ⓨ from being a binary operation.

Exercise 6–27.   Let & stand for the combining rule:

$$a \& b = \text{the average of } a \text{ and } b \text{, that is, the number } \frac{a + b}{2} \text{ that}$$
$$\text{is halfway between } a \text{ and } b.$$

Show that & is not a binary operation on the integers, because the average is not always an integer. However, the rule & *is* a binary operation for the rational numbers (all possible fractions). Can you prove that it is?

Exercise 6–28.   Take as elements the people now in this building.

Define a combining rule # as follows:

$$a \mathbin{\#} b = \begin{cases} \text{the female if one of } a, b \text{ is male and the other} \\ \quad \text{the other female} \\ \text{the shorter of } a, b \text{ if both are male} \end{cases}$$

Under what circumstances would # fail to be a binary operation? Suppose that at 2 a.m. one switchboard operator is the only female in the building? Will # automatically be a binary operation, or could it still fail?

Exercise 6–29. Show that composition ∘ is a binary operation on the 6 job exchanges described at the beginning of this chapter.

Exercise 6–30. Consider the 6 rotations and flips of an equilateral triangle in Figure 6–3. The combining rule composition ∘ calls for performing rotations and flips in sequence. Show that composition is a binary operation in this case.

Exercise 6–31. Let the rule (*A.P.*) be defined on the integers so that for each two integers i and j, $i\,(A.P.)\,j$ is the next integer in the arithmetic progression $i, j, \_\_, \dots.$ For instance, $3\,(A.P.)\,5 = 7$; $2\,(A.P.)\,8 = 14$; $99\,(A.P.)\,88 = 77$. Find a formula for $i\,(A.P.)\,j$, and show that (*A.P.*) is a binary operation.

Exercise 6–32. Show that the join ∨ and meet ∧ of Definition 4–1 are binary operations.

Exercise 6–33. Show that "or ∨" and "and ∧" are binary operations on all the logical statements in a space.

Exercise 6–34. Show that union ∪ and intersection ∩ are binary operations on all the subsets of a space.

Systems showing some of the same features as the 6 job exchanges or symmetries of a triangle appear repeatedly in mathematics and have a name of their own:

Definition 6–2. A **group** is a mathematical system of elements with a binary operation ∗ having three properties,

   i. the **associative** property. For each 3 elements a, b, c,

$$(a \ast b) \ast c = a \ast (b \ast c).$$

   ii. the **identity** property. There is an identity element e for which

$$a \ast e = e \ast a = a, \text{ for all } a.$$

iii. the **inverse** property. For each element a, there is an inverse element a^{-1} for which

$$a * a^{-1} = a^{-1} * a = e.$$

A familiar feature of arithmetic is missing in many groups, the property of commutativity. If $a * b = b * a$, then a and b are said to **commute** under the operation $*$.

A **commutative group** (or **abelian group**, honoring N. H. Abel [1802–1829]) is a group in which every two elements commute under the group operation.

Definition 6–2 describes the system mainly through its behavior. It does not detail just what the "elements" are, but it does tell how they interact when combined by the binary operation $*$. Such definitions through behavior are typical of modern approaches to mathematics. One major advantage is versatility. Since the elements of the group are left unspecified, they can be drawn from a great variety of examples.

EXAMPLES   A point-by-point check shows that the 6 job exchanges form a group: From Exercise 6–29, composition is a binary operation.

i. The 6 exchanges have the associative property under $\circ$, since both $(a \circ b) \circ c$ and $a \circ (b \circ c)$ stand for the sequence

exchange c—followed by exchange b—followed by exchange a.

ii. The identity exchange is the trivial exchange

$$e = \begin{pmatrix} X & Y & Z \\ X & Y & Z \end{pmatrix},$$

as shown in Exercise 6–16.

iii. The inverses are

$a^{-1} = ?$ (Check that $a \circ d = d \circ a = e$, so that $a^{-1} = d$.)

$b^{-1} = ?$ (Exercise 6–12. Show that $b^{-1} = b$.)

$c^{-1} = ?$ (Exercise 6–13.)

$d^{-1} = a$

$e^{-1} = e$

$f^{-1} = f$ (Exercise 6–14.)

The group of 6 job exchanges is not a commutative group, since, for instance, $a \circ b = c$, but $b \circ a = f$. As in every group, however, the identity e commutes with every element.

The counting numbers $1, 2, 3, \ldots$ do not form a group under addition. Why? Addition is a binary operation, for the sum of any two counting numbers is exactly one counting number.

i. The system has the associative property, $(m + n) + q = m + (n + q)$.

ii. There is no identity element in the system, for $m + n$ is greater than m, even if n is *1*.

iii. The inverse of each element is defined in terms of the identity, so the failure of property *ii* implies the failure of property *iii*, too.

Now consider the non-negative integers 0, 1, 2, 3, . . . under addition. Again, addition is a binary operation, and (*i*) the system is associative. (*ii*) The identity element is 0, for $n + 0 = 0 + n = n$, for each number n. (*iii*) The system does *not* have the inverse property, for there is no inverse for 1 that can be added to 1 to give zero. (Recall that -1 is not included, since it is a negative integer.) Notice that 2 does not have an additive inverse, either, nor does 3, nor 4, and so forth. In fact, the only element that has an additive inverse is 0: $0 + 0 = 0$.

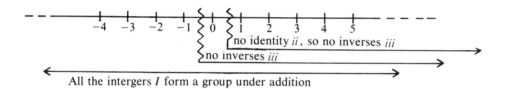

Exercise 6–35. Check point by point that all the integers, positive, negative, and zero, do form a group under addition. Note that the group is commutative.

Exercise 6–36. Show that the integers, 0, 1, -1, 2, -2, . . . under multiplication satisfy all but one of the group requirements,

Exercise 6–37. Show that the non-zero fractions n/d, $n \neq 0$, $d \neq 0$, form a group under multiplication. (Note that the multiplicative inverse of $-129/22$, say, is $-22/129$, since $[-129/22] \times [-22/129] = 1$. What is the product $[n/d] \times [d/n]$?) Is the group commutative?

In Exercises 6–35 and 6–37 you have learned about two important groups, the integers with $+$, and the non-zero fractions with $\times$. As you have seen, the non-negative integers with $+$ fail to be a group, because they do not have the inverse property. This, in fact, is the reason for introducing negative numbers; they provide additive inverses for positive numbers. Similarly, the non-zero integers with $\times$ lack inverses, which can be supplied by introducing fractions.

Modern methods of teaching arithmetic emphasize the group properties instead of obscuring them. Subtraction, which is not even an associative operation on the integers ($9 - [5 - 2] \neq [9 - 5] - 2$, for instance), is de-emphasized, appearing in its group role as inverse to addition. A problem that would once have been presented as $15 - 8 = \square$ appears as an (inverse) addition problem $8 + \square = 15$.

In the following exercises you will learn that groups arise in varied contexts.

Exercise 6-38. You have an electrical switch with 2 settings, open and closed. Let s stand for switching the switch; that is, changing it from whatever its initial setting is to the other setting. Let 0 stand for leaving the switch alone. Then under composition ∘, that is, switchings performed in sequence, 0 and s combine as shown in this table:

∘	0	s
0	0	s
s	s	0

The table gives $0 \circ 0 = 0$, $0 \circ s = s$, $s \circ 0 = s$, and $s \circ s = 0$. Show that $s^3 = s \circ s \circ s = s$. Show that $\{0, s\}$, ∘ is a commutative group.

Exercise 6-39. Let e stand for "even integer" and u stand for "odd (uneven) integer" and let the binary operation on e and u be addition. Then show that the addition table is

+	e	u
e	e	u
u	u	e

Show that e and u with + form a group.

Exercise 6-40. For the integers taken modulo 2 (Chapter 2) the addition table is given by

+	0	1
0	0	1
1	1	0

(mod 2)

Show that the integers modulo 2 with + form a commutative group.

The groups of Exercises 6-38, 6-39, and 6-40 are alike except for names and notation. Mathematicians say that they are "abstractly identical" or "isomorphic," meaning that as far as their group properties are concerned they are just alike. A group that is abstractly identical to the 2-membered group of Exercise 6-40 is called a **cyclic group of order 2.** It gets its name from the fact that it has 2 elements that occur in cycles as the operation is repeated on the nonidentity element: For instance, in

Exercise 6–39,

$$u = u$$

$$u + u = e$$

$$u + u + u = u$$

$$u + u + u + u = e,$$

and so forth.

Exercise 6–41. You have two electrical switches, a red one and a green one. Let a stand for switching the red switch and let b stand for switching the green switch. As in Exercise 6–38, the initial position of each switch does not matter; it is the change or lack of change that interests us. Let c stand for switching both switches, and let e stand for leaving both switches alone. Taking composition ∘ as the binary operation, the switchings combine as in this Cayley table:

∘	e	a	b	c
e	e	a	b	c
a	a	e	c	b
b	b	c	e	a
c	c	b	a	e

Describe what composition means in terms of the switches, and so explain why the system has the associative property. Explain why e is the identity. Explain why each element is its own inverse, drawing on Exercise 6–38 for comparison. Put these observations together to prove that the system forms a group.

The commutative group of Exercise 6–41 is called the **Klein four-group** (for Felix Klein [1849–1925]). The elements can be given different names and the binary operation can be changed, but all groups of 4 elements, each its own inverse, are the same as far as their group arithmetic is concerned. They all have Cayley tables like that of Exercise 6–41, except for the names of the elements, and so they are all called the Klein four-group.

Exercise 6–42. The compound statement $P \wedge Q$ stands for *P and Q* (Chapter 3). Let a stand for negating the first part, P, and let b stand for negating Q. Let c stand for negating both parts. Develop from this different setting the Cayley table of the Klein four-group, as in Exercise 6–41.

Exercise 6–43. Let 25 stand for an amount of money that is some integral number of dollars plus one quarter in change. Similarly, let 50 represent some dollars plus two quarters in change, 75 three quarters in change. With binary operation +, derive the Cayley table

+	0	25	50	75
0	0	25	50	75
25	25	50	75	0
50	50	75	0	25
75	75	0	25	50

Show that the system is a commutative group.

Exercise 6–44. Let 15 represent a length of time 15 minutes more than an integral number of hours. Let 30 represent an interval of half an hour plus some hours, 45 three-quarters of an hour plus some hours. By taking times consecutively, develop the Cayley table

+	0	15	30	45
0	0	15	30	45
15	15	30	45	0
30	30	45	0	15
45	45	0	15	30

Show that the system is a commutative group.

Exercise 6–45. Write the addition table for the integers taken modulo 4 (Chapter 2), and show that the integers (mod 4) with + constitute a commutative group.

The groups of Exercises 6–43, 6–44, and 6–45 look different notationally, but, as you may have noticed already, they are abstractly alike. Such groups are called **cyclic groups of order 4.** The 4 elements occur in cycles as the operation is repeated on an appropriate element. For instance, in Exercise 6–44,

$$15 = 15$$
$$15 + 15 = 30$$
$$15 + 15 + 15 = 45$$
$$15 + 15 + 15 + 15 = 0 \,(+1 \text{ hr}) \equiv 0)$$
$$15 + 15 + 15 + 15 + 15 \equiv 15$$
$$15 + 15 + 15 + 15 + 15 + 15 \equiv 30,$$

$$45 = 45$$
$$45 + 45 \equiv 30$$
$$45 + 45 + 45 \equiv 15$$
$$45 + 45 + 45 + 45 \equiv 0$$
$$45 + 45 + 45 + 45 + 45 \equiv 45$$
$$45 + 45 + 45 + 45 + 45 + 45 \equiv 30,$$

and so forth.

In this way, 15 or 45 generates the whole cycle of four elements, but 0 and

30 do not, since $0 + \cdots + 0 = 0$, and $30 + 30 \equiv 0$, $30 + 30 + 30 \equiv 30$, and so on.

Exercise 6–46. Take as elements the non-zero integers modulo 5, in the order 1, 2, 4, 3. Take multiplication modulo 5 as the binary operation. Construct the Cayley table and show that the system is a group, the cyclic group of order 4.

Exercise 6–47. Referring to Chapter 2, fill in this outline of a proof that the integers with addition taken modulo m form a group: Associativity of the integers as a whole implies associativity of the integers modulo m. The identity is the 0 class of multiples of m. What class forms the additive inverse for integers of the form $a + km$?

The Octic Group of Symmetries of a Square

Cut a square of paper, shade one side, and number the corners as in Figure 6–4. There are 8 different ways the square can be lifted and replaced

FIGURE 6–4. PAPER MODEL FOR STUDYING SYMMETRIES OF A SQUARE.

in the same space. Four of these symmetries rotate the square without turning it over to the unshaded side. We shall call the 4 rotations of the square r, s, t, and e.

$$e = \begin{pmatrix} 1 & 2 & 3 & 4 \\ 1 & 2 & 3 & 4 \end{pmatrix}, \quad r = \begin{pmatrix} 1 & 2 & 3 & 4 \\ 4 & 1 & 2 & 3 \end{pmatrix},$$

$$s = \begin{pmatrix} 1 & 2 & 3 & 4 \\ 3 & 4 & 1 & 2 \end{pmatrix}, \quad t = \begin{pmatrix} 1 & 2 & 3 & 4 \\ 2 & 3 & 4 & 1 \end{pmatrix}.$$

You recognize e as the identity symmetry that replaces the square exactly as it was. In symmetry r, vertex 1 is replaced by vertex 4, vertex 2 by 1, 3 by 2, and 4 by 3.

Exercise 6–48. Experiment with a paper square to show that r rotates the square 90°, that s is a rotation of 180°, t a rotation of 270°, e a rotation of 0°.

The other four symmetries are flips, that is, they turn the square over from the shaded side to the unshaded side, or from the unshaded side to the shaded side:

$$h = \begin{pmatrix} 1 & 2 & 3 & 4 \\ 4 & 3 & 2 & 1 \end{pmatrix}, \quad v = \begin{pmatrix} 1 & 2 & 3 & 4 \\ 2 & 1 & 4 & 3 \end{pmatrix},$$

$$d = \begin{pmatrix} 1 & 2 & 3 & 4 \\ 1 & 4 & 3 & 2 \end{pmatrix}, \quad a = \begin{pmatrix} 1 & 2 & 3 & 4 \\ 3 & 2 & 1 & 4 \end{pmatrix}.$$

Exercise 6–49. Demonstrate each of the 4 flips by means of a paper square. Demonstrate that if you perform the same flip twice in sequence, the square returns to its initial position.

EXAMPLE As you demonstrated in Exercise 6–49, a square flipped twice in sequence by the same flip returns to its initial position. Using composition to combine v and v, we have

$$v \circ v = \begin{pmatrix} 1 & 2 & 3 & 4 \\ 2 & 1 & 4 & 3 \end{pmatrix} \circ \begin{pmatrix} 1 & 2 & 3 & 4 \\ 2 & 1 & 4 & 3 \end{pmatrix} = \begin{pmatrix} 1 & 2 & 3 & 4 \\ 1 & 2 & 3 & 4 \end{pmatrix} = e,$$

the identity transformation.

Exercise 6–50. Find $h \circ h$, $d \circ d$, and $a \circ a$.

To find other combinations of symmetries, experiment with the paper square and form the composition of the symmetries.

EXAMPLE To find the net effect of $h \circ r$ on the square, first rotate the paper square 90° so that 1 is replaced by 4, 2 by 1, 3 by 2, and 4 by 3. Then flip the square so as to interchange 1 and 4 and interchange 2 and 3 (Figure 6–5). The net effect on the square is to flip it, keeping vertices 1 and 3 fixed and interchanging 2 and 4, that is, the flip d. We can write the composition in

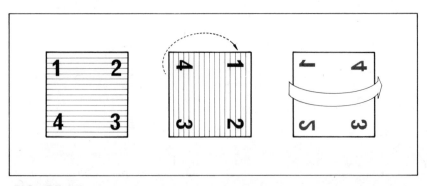

FIGURE 6–5. THE COMPOSITION $h \circ r$ DEMONSTRATED WITH A PAPER SQUARE.

terms of the vertices:

$$h \circ r = \begin{pmatrix} 1 & 2 & 3 & 4 \\ 4 & 3 & 2 & 1 \end{pmatrix} \circ \begin{pmatrix} 1 & 2 & 3 & 4 \\ 4 & 1 & 2 & 3 \end{pmatrix} = \begin{pmatrix} 1 & 2 & 3 & 4 \\ 1 & 4 & 3 & 2 \end{pmatrix} = d.$$

Exercise 6–51. Show that $r \circ h = a$. Demonstrate with a paper square that $h \circ r \neq r \circ h$, the final position depending on whether the flip precedes or follows the rotation. Then these two symmetries do not commute.

Exercise 6–52. Show that $t \circ h = d$ and $h \circ t = a$. Check with a paper square that the final position does indeed vary with the order.

Exercise 6–53. Show that $r \circ r = s$, $r \circ r \circ r = r \circ s = t$, and $r \circ r \circ r \circ r = r \circ t = e$. Interpret this for a paper square.

Exercise 6–54. Show that $t \circ t = s$, $t \circ t \circ t = t \circ s = r$, and $t \circ t \circ t \circ t = t \circ r = e$. Interpret this for a paper square.

Exercise 6–55. Find five symmetries besides e that bring the square back to original position when performed twice in succession. (For example, $d \circ d = e$.)

Exercise 6–56. Four prisoners, whose convict numbers end in 1, 2, 3, and 4, respectively, exchange prison jobs according to exchanges r and h. Show that by applying different sequences of these two exchanges, they will accomplish as net exchanges the eight possible symmetries of a square.

Figure 6–6 presents all the compositions of the **octic group** ("octic" means "8") of the symmetries of a square.

$\circ$	e	r	s	t	h	v	d	a
e	e	r	s	t	h	v	d	a
r	r	s	t	e	a	d	h	v
s	s	t	e	r	v	h	a	d
t	t	e	r	s	d	a	v	h
h	h	d	v	a	e	s	r	t
v	v	a	h	d	s	e	t	r
d	d	v	a	h	t	r	e	s
a	a	h	d	v	r	t	s	e

FIGURE 6–6. CAYLEY TABLE OF THE OCTIC GROUP.

Exercise 6–57. Define 8 matrices (Chapter 5) as follows:

$$E = \begin{pmatrix} 1 & 0 \\ 0 & 1 \end{pmatrix}, \quad R = \begin{pmatrix} 0 & -1 \\ 1 & 0 \end{pmatrix}, \quad S = \begin{pmatrix} -1 & 0 \\ 0 & -1 \end{pmatrix}, \quad T = \begin{pmatrix} 0 & 1 \\ -1 & 0 \end{pmatrix},$$

$$H = \begin{pmatrix} 0 & 1 \\ 1 & 0 \end{pmatrix}, \quad V = \begin{pmatrix} 0 & -1 \\ -1 & 0 \end{pmatrix}, \quad D = \begin{pmatrix} 1 & 0 \\ 0 & -1 \end{pmatrix}, \quad A = \begin{pmatrix} -1 & 0 \\ 0 & 1 \end{pmatrix}.$$

Construct a multiplication table, showing that the 8 matrices form a group with matric multiplication as the binary operation. (Associativity can be proved or assumed.) The Cayley table coincides (except for the capital letters) with that of the octic group shown in Figure 6–6, but notice that the elements are matrices, not symmetries or job exchanges, so that *RH*, for instance, means the matric product having *R* on the left:

$$RH = \begin{pmatrix} 0 & -1 \\ 1 & 0 \end{pmatrix} \begin{pmatrix} 0 & 1 \\ 1 & 0 \end{pmatrix}.$$

PERMUTATIONS

Exchanges such as $\begin{pmatrix} X & Y & Z \\ Z & Y & X \end{pmatrix}$ and $\begin{pmatrix} 1 & 2 & 3 & 4 \\ 4 & 3 & 2 & 1 \end{pmatrix}$ are called *per-mutations*. A **permutation P** on a set S shows which element t of S replaces each element s of S, by the notation

$$\begin{pmatrix} - & - & - & s & - & - & - \\ - & - & - & t & - & - & - \end{pmatrix},$$

with t written directly under the element s that it replaces. Since each s in S is replaced by exactly one element t, both rows of the permutation consist of the elements of S in some order.

EXAMPLES In trying to dial a telephone number, 123–4567, I mistakenly dial the digits out of order as 213–5746. I have *permuted* the digits, according to the *permutation*

$$\begin{pmatrix} 1 & 2 & 3 & 4 & 5 & 6 & 7 \\ 2 & 1 & 3 & 5 & 7 & 4 & 6 \end{pmatrix}.$$

A small boy says that the last three digits on the license plate of a getaway car were 4 and 8 and 6, but he cannot remember the order. The Motor Vehicle Department checks its files for all six *permutations;* 468, 486, 648, 684, 846, and 864.

The dealer shuffles a pack of 52 bridge cards. The shuffled cards represent a *permutation* of the original order.

There are exactly 6 ways to arrange three elements, X, Y, Z: If X is first, either Y or Z can be second, giving the 2 orders

$$X\,Y\,Z \quad \text{and} \quad X\,Z\,Y.$$

Similarly, there are 2 orders having Y first, and 2 orders having Z first.

EXAMPLE One of the 6 orders of the set X, Y, Z is $Y\,X\,Z$. If the letters are taken alphabetically for the first row, and in the order $Y\,X\,Z$ for the second row, the resulting permutation is

$$\begin{pmatrix} X & Y & Z \\ Y & X & Z \end{pmatrix},$$

which was designated c in Figure 6–2.

Exercise 6–58. Write the 6 orders of three elements X, Y, and Z. Form the 6 permutations on the set $\{X, Y, Z\}$, and identify them by the letters used in Figure 6–2.

The symmetries of a triangle include all 6 permutations of the vertices, but the 8 symmetries of a square do not include all permutations of the vertices, since some permutations would require twisting the square out of shape, as shown in Figure 6–7.

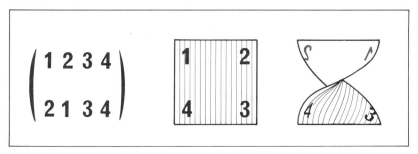

FIGURE 6–7. THE PERMUTATION $\begin{pmatrix} 1 & 2 & 3 & 4 \\ 2 & 1 & 3 & 4 \end{pmatrix}$ ACTING ON A SQUARE DEFORMS THE SQUARE.

How many possible arrangements are there for the four elements 1, 2, 3, 4? Suppose 3 is in the first position. Then the other positions are filled with the three other elements, 1, 2, and 4 in some order. There are $3 \cdot 2 = 6$ ways to permute three elements. Now suppose 2 is in the first position. Again there are $3 \cdot 2 = 6$ corresponding arrangements for 1, 3, and 4, yielding 6 permutations

$$\begin{pmatrix} 1 & 2 & 3 & 4 \\ 2 & 1 & 3 & 4 \end{pmatrix}, \quad \begin{pmatrix} 1 & 2 & 3 & 4 \\ 2 & 1 & 4 & 3 \end{pmatrix}, \quad \begin{pmatrix} 1 & 2 & 3 & 4 \\ 2 & 3 & 1 & 4 \end{pmatrix},$$

$$\begin{pmatrix} 1 & 2 & 3 & 4 \\ 2 & 3 & 4 & 1 \end{pmatrix}, \quad \begin{pmatrix} 1 & 2 & 3 & 4 \\ 2 & 4 & 1 & 3 \end{pmatrix}, \quad \begin{pmatrix} 1 & 2 & 3 & 4 \\ 2 & 4 & 3 & 1 \end{pmatrix}.$$

Exercise 6–59. Write the $4 \cdot 3 \cdot 2 \cdot 1 = 24$ arrangements of the elements 1, 2, 3, 4. Write the 6 permutations that hold two of the four elements fixed and interchange the other two. Which of these 6 appear in the octic group?

The symbol $n!$ is read **"n factorial"** and stands for

$$n! = n(n-1)(n-2) \cdots 3 \cdot 2 \cdot 1.$$

For instance, $1! = 1$, $2! = 2 \cdot 1 = 2$, $3! = 3 \cdot 2 \cdot 1 = 6$, $4! = 4 \cdot 3! = 24$, $5! = 5 \cdot 4! = 5 \cdot 4 \cdot 3 \cdot 2 \cdot 1 = 120$, and so on.

Exercise 6–60. You have seen that there are 4! permutations on 4 elements, because for each of the 4 choices for the first position, the remaining 3 elements have 3! arrangements. Explain why there are $n!$ permutations on n elements.

The $n!$ permutations on n elements, with composition, form a group, as we can prove by checking the group requirements one at a time.

First, composition is a binary operation on the permutations, for if P and Q are two permutations on n elements, then the composition $P \circ Q$ is the permutation that replaces each element s by u, where Q replaces s by t and P replaces t by u:

$$P \circ Q = \begin{pmatrix} --- t --- \\ --- u --- \end{pmatrix} \circ \begin{pmatrix} --- s --- \\ --- t --- \end{pmatrix} = \begin{pmatrix} --- s --- \\ --- u --- \end{pmatrix}.$$

There is no ambiguity about the answer permutation $P \circ Q$; it is exactly one of the $n!$ permutations.

EXAMPLE

For $n = 6$, let P and Q be defined by

$$P = \begin{pmatrix} 1 & 2 & 3 & 4 & 5 & 6 \\ 2 & 4 & 6 & 1 & 3 & 5 \end{pmatrix} \quad \text{and} \quad Q = \begin{pmatrix} 1 & 2 & 3 & 4 & 5 & 6 \\ 6 & 5 & 4 & 3 & 2 & 1 \end{pmatrix}.$$

Then

$$P \circ Q = \begin{pmatrix} 1 & 2 & 3 & 4 & 5 & 6 \\ & & & & & \\ 2 & 4 & 6 & 1 & 3 & 5 \end{pmatrix} \circ \begin{pmatrix} 1 & 2 & 3 & 4 & 5 & 6 \\ & & & & & \\ 6 & 5 & 4 & 3 & 2 & 1 \end{pmatrix}$$

$$= \begin{pmatrix} 1 & 2 & 3 & 4 & 5 & 6 \\ 5 & 3 & 1 & 6 & 4 & 2 \end{pmatrix}.$$

If P, Q, and R are three permutations on n elements, then $(P \circ Q) \circ R = P \circ (Q \circ R)$, because both stand for the same sequence of changes, R followed by Q followed by P. Therefore, the system is associative.

The identity permutation replaces each element s by itself:

$$e = \begin{pmatrix} --- s --- \\ --- s --- \end{pmatrix}.$$

The inverse of a permutation that replaces s by t replaces t by s:

$$\text{If} \quad P = \begin{pmatrix} --- s --- \\ --- t --- \end{pmatrix}, \quad \text{then} \quad P^{-1} = \begin{pmatrix} --- t --- \\ --- s --- \end{pmatrix}. \quad \blacksquare$$

For $n = 6$, let $P = \begin{pmatrix} 1 & 2 & 3 & 4 & 5 & 6 \\ 2 & 4 & 6 & 1 & 3 & 5 \end{pmatrix}$. Then

$$P^{-1} = \begin{pmatrix} 2 & 4 & 6 & 1 & 3 & 5 \\ 1 & 2 & 3 & 4 & 5 & 6 \end{pmatrix},$$

which can be rearranged to put the first row in counting order for easier reference,

$$P^{-1} = \begin{pmatrix} 1 & 2 & 3 & 4 & 5 & 6 \\ 4 & 1 & 5 & 2 & 6 & 3 \end{pmatrix},$$

for their composition is the identity permutation e:

$$P \circ P^{-1} = \begin{pmatrix} 1 & 2 & 3 & 4 & 5 & 6 \\ 2 & 4 & 6 & 1 & 3 & 5 \end{pmatrix} \circ \begin{pmatrix} 1 & 2 & 3 & 4 & 5 & 6 \\ 4 & 1 & 5 & 2 & 6 & 3 \end{pmatrix}$$

$$= \begin{pmatrix} 1 & 2 & 3 & 4 & 5 & 6 \\ 1 & 2 & 3 & 4 & 5 & 6 \end{pmatrix} = e.$$

In the next four exercises, let $n = 5$, and let permutations P, Q, and R be defined:

$$P = \begin{pmatrix} 1 & 2 & 3 & 4 & 5 \\ 2 & 3 & 1 & 5 & 4 \end{pmatrix}, \qquad Q = \begin{pmatrix} 1 & 2 & 3 & 4 & 5 \\ 2 & 1 & 3 & 5 & 4 \end{pmatrix},$$

$$R = \begin{pmatrix} 1 & 2 & 3 & 4 & 5 \\ 3 & 2 & 1 & 5 & 4 \end{pmatrix}.$$

Exercise 6-61. Find $P \circ Q$ and $Q \circ R$.

Exercise 6-62. Find $(P \circ Q) \circ R$ and $P \circ (Q \circ R)$. Verify associativity in this case.

Exercise 6-63. Find P^{-1}, $P^{-1} \circ P$, Q^{-1}, and $Q^{-1} \circ Q$. Find $Q \circ Q^{-1}$.

Exercise 6-64. Find $Q^{-1} \circ P^{-1}$. Show that $(Q^{-1} \circ P^{-1}) \circ (P \circ Q) = e$.

You have been introduced to groups through groups of permutations. Now you see that all $n!$ permutations of any n elements form a group with operation composition.

It turns out that the connection between groups and permutations is even closer. According to "Cauchy's Representation Theorem for Groups,"[*] every group can be represented as a group of permutations. The method is based on a feature of the Cayley table which we will now explore.

---

[*] See, for instance, J. E. and M. W. Maxfield, *Abstract Algebra and Solution by Radicals*, W. B. Saunders Company, 1971, p. 58.

THE CAYLEY TABLE OF A GROUP

You have been using Cayley tables to record how elements combine under the group operation. Several features of the group are immediately apparent from the table. Since the combining rule is an operation, there is one and only one entry in each row and each column, and each entry is a group element. This reflects the fact that the entry in the a row, b column, $a * b$, is exactly one group element. From the identity property, there must be an e row that merely repeats the column headings ($e * a = a$ for each a) and an e column that repeats the row headings ($a * e = a$ for each a). From the inverse property, the identity e appears in every row ($a * a^{-1} = e$) and in every column ($a^{-1} * a = e$).

Not only e, but *every* group element appears in every row and in every column. To see why, consider the Cayley table of the octic group in Figure 6–6. To show why every element appears in, say, the t row, pick an element, say a, and predict where it can be found in the t row. First, use the table to find $t^{-1} = r$. Then find $t^{-1} \circ a = r \circ a = v$ from the table. Then a must appear in the v column of the t row, since

$$t \circ v = t \circ (r \circ a) = (t \circ r) \circ a = e \circ a = a.$$

In fact, each group element appears exactly once in each row and in each column of the Cayley table. Suppose a occurred elsewhere in the t row, say in the h column. Then from $t \circ v = a$ and $t \circ h = a$,

$$t \circ v = t \circ h$$
$$r \circ (t \circ v) = r \circ (t \circ h)$$
$$(r \circ t) \circ v = (r \circ t) \circ h$$
$$e \circ v = e \circ h$$
$$v = h$$

The method illustrated here for one row of the octic group table is perfectly general and can be used to prove that **each row and each column of a group table is an arrangement of the group elements in some order.** In this way, combining the group elements one at a time with one of the elements produces a permutation of the elements. For example, in the four group, combining the elements one at a time with the element b to get the b row of the table,

$$b * e, \quad b * a, \quad b * b, \quad b * c,$$

produces the permutation of the group elements

$$b \sim \begin{pmatrix} e & a & b & c \\ b & c & e & a \end{pmatrix}.$$

Exercise 6–65. Verify that each group element occurs exactly once in each row and in each column in some of the Cayley tables you have studied.

SUBGROUPS

Sometimes you may notice in the Cayley table of a group a block of elements that form a group of their own within the original group. A group H is called a **subgroup** of a group G if they have the same operation $*$ and if all the elements of H are elements of G.

In Figure 6–6 notice that the upper left quarter of the octic group table **EXAMPLES** contains only the four elements $e, r, s,$ and t. Taking that quarter by itself, you have the Cayley table for a subgroup H of the octic group.

Exercise 6–66. Verify that the elements $e, r, s,$ and t of the octic group form a group: Does each combination $r \circ t$, $s \circ r$, $e \circ e$, and so on produce exactly one answer, $e, r, s,$ or t? Explain how you know that $r \circ (r \circ t) = (r \circ r) \circ t$, for instance, just because each element of H is an element of G, the octic group. Which is the identity element? List the inverse of each element.

The octic group has other subgroups, for instance, $V : \{e, v\}$, $A : \{e, a\}$, $S : \{e, s\}$. Another of its subgroups is $E : \{e\}$, the group having just the single element e.

Exercise 6–67. Follow the outline in Exercise 6–66 to show that V is a subgroup of the octic group.

Exercise 6–68. Prove that every group has the "trivial" subgroup $E : \{e\}$, having the group identity as its only element.

Exercise 6–69. Prove that if G is a group, then G is a subgroup of G, since it satisfies the definition, if only in a rather technical way.

The subgroups G and E of a group G are called **trivial** subgroups. Subgroups can have an interesting interpretation depending on the original source of the group. As you have seen, the octic group can be derived from the symmetries of a square. Consider its subgroup $H : \{e, r, s, t\}$. The elements of H are those derived from rotations of the square, that is, from symmetries that do not involve flipping the square to the other side. The restriction in the octic group to just the four elements of the subgroup has a corresponding restriction in the physical model. This suggests why subgroups are important in the analysis of a group.

The integers I form a group with operation $+$. Let $\langle 6 \rangle$ stand for all **EXAMPLE** integers that are multiples of 6, with operation $+$. Then $\langle 6 \rangle$ is a subgroup of I: Let $6i$ and $6j$ be any two elements of $\langle 6 \rangle$. Then $6i + 6j = 6(i + j) = 6k$, where k is the sum integer of i and j. Since $6k$ is a multiple of 6, the sum $6i + 6j$ is exactly one element of $\langle 6 \rangle$. The associative property in $\langle 6 \rangle$ is inherited from I, since three integers that are multiples of 6 are also elements of I. The identity of $\langle 6 \rangle$ is 0, the identity of I. The additive inverse of $6i$ is $6(-i)$, again a multiple of 6.

Let $\langle 2 \rangle$ represent even integers, multiples of 2, with operation $+$. Then $\langle 2 \rangle$ is a subgroup of I, and also $\langle 6 \rangle$ is a subgroup of $\langle 2 \rangle$.

Exercise 6–70. Show that $\langle 10 \rangle$ is a subgroup of I and that $\langle 10 \rangle$ is a subgroup of $\langle 5 \rangle$. Suppose you are operating an adding machine. How do you punch an element of $\langle 10 \rangle$? Suppose that one page of addition (and subtraction) involves only elements of $\langle 10 \rangle$. Will any of the answers have a non-zero units digit? Do you need to carry a units digit for that page? Suppose you are making change in a store for items costing multiples of 5¢. Assuming no tax to be computed, what is the lowest denomination coin you need for making change? Suppose each item costs a multiple of 10¢. Then what is the lowest denomination coin you need?

Exercise 6–71. Let M be the group of all 2 by 2 matrices of integers, with matric addition as the binary operation (Chapter 5). Let N represent the 2 by 2 matrices of integers having a zero in the 1, 1 position—$\begin{pmatrix} 0 & 1 \\ 2 & 3 \end{pmatrix}$, for instance. Show that N is a subgroup of M.

Exercise 6–72. The nonzero rational numbers, such as $-\frac{2}{3}$, $\frac{71}{1000}$, for instance, form a group with multiplication as operation. Show that the two elements $+1$ and -1 form a subgroup.

RINGS AND FIELDS

You have found that the integers I have one binary operation, $+$, under which they form a commutative group, and a second binary operation, $\times$, under which they are not a group. The integers offer an important example of a *ring*.

> **Definition 6–3.** A mathematical system R with two binary operations, here called $+$ and $\times$ to suggest operations on integers, is a **ring** if
>
>   *i.* The system with operation $+$ forms a commutative group.
>
>   *ii.* The operation $\times$ is associative.
>
>   *iii.* The operation $\times$ is **distributive** with respect to $+$
>
>   on the left: $a \times (b + c) = (a \times b) + (a \times c)$,
>
>   and on the right: $(b + c) \times a = (b \times a) + (c \times a)$,
>
>   for all elements a, b, c in R.

Some rings have a multiplicative identity e for which $a \times e = e \times a$ for each a in the ring, and so are called **rings with unit element.** For example, the integers I have 1 as a multiplicative identity.

The integers have the additional property that multiplication is commutative, and so form a **commutative ring with unit element**

The integers fail to be a group under multiplication, since most integers lack multiplicative inverses. (For example, there is no integer i for which $3i = i3 = 1$.) The failure of the inverse property reflects practical problems: For instance, suppose you have 15 cookies to deal evenly to 6 children. If you give each child 2 cookies, you distribute only $6 \times 2 = 12$, but you do not have enough for 3 apiece. If you can break cookies into fractions so as to give each child $2\frac{1}{2}$, you can solve the problem, since $6 \times 2\frac{1}{2} = 15$.

To solve the problem of multiplicative inverses for integers, we form the rational numbers, made up of all quotients n/d, where n and d are integers, d non-zero. The rational numbers form a field.

Definition 6–4. A **field** is a mathematical system F of at least 2 elements, with binary operations $+$ and $\times$, such that

i. F is a commutative group under $+$.

ii. The non-zero elements of F form a commutative group under $\times$.

iii. $\times$ is distributive with respect to $+$.

Alternatively, a field can also be defined as a commutative ring with unit element all of whose non-zero elements have multiplicative inverses.

Let M be the 2 by 2 matrices of integers (Chapter 5). Then M has a **EXAMPLE** binary operation $+$, under which it is a commutative group.

The additive identity is the zero matrix

$$\begin{pmatrix} 0 & 0 \\ 0 & 0 \end{pmatrix}.$$

The additive inverse of

$$\begin{pmatrix} 1 & 2 \\ -3 & 0 \end{pmatrix} \text{ is } \begin{pmatrix} -1 & -2 \\ 3 & 0 \end{pmatrix}.$$

There is a second binary operation defined on M, matric multiplication, which is associative and which distributes on the left and on the right with respect to matric addition, as demonstrated in Chapter 5. Matric multiplication is not commutative.

$$\begin{pmatrix} 0 & 1 \\ 1 & 0 \end{pmatrix}\begin{pmatrix} 2 & 3 \\ 0 & 4 \end{pmatrix} = \begin{pmatrix} 0 & 4 \\ 2 & 3 \end{pmatrix}, \text{ but } \begin{pmatrix} 2 & 3 \\ 0 & 4 \end{pmatrix}\begin{pmatrix} 0 & 1 \\ 1 & 0 \end{pmatrix} = \begin{pmatrix} 3 & 2 \\ 4 & 0 \end{pmatrix},$$

for instance. There is a multiplicative identity in M,

$$\begin{pmatrix} 1 & 0 \\ 0 & 1 \end{pmatrix},$$

but no multiplicative inverse for such matrices as

$$\begin{pmatrix} 1 & 0 \\ 0 & 0 \end{pmatrix},$$

for instance, as you could show by multiplying by a matrix

$$\begin{pmatrix} a & b \\ c & d \end{pmatrix}$$

and showing that no choice of a and c could make the product,

$$\begin{pmatrix} a & 0 \\ c & 0 \end{pmatrix},$$

equal the identity matrix. The 2 by 2 matrices over the ring of integers form a ring with unit element. They are not a commutative ring and they are not a field.

Exercise 6–73. Which matrix has a multiplicative inverse,

$$\begin{pmatrix} 2 & 4 \\ 1 & 2 \end{pmatrix} \quad \text{or} \quad \begin{pmatrix} 3 & 2 \\ 1 & 1 \end{pmatrix}?$$

(See page 149 in Chapter 5 for inverses of matrices.)

EXAMPLE Let $I/\langle 6 \rangle$ stand for the six distinct equivalence classes modulo 6 (Chapter 2). Since $6 \equiv 0, 27 \equiv 3$, and in general $a + b6 \equiv a \pmod 6$, the sums and products can be given compactly in these tables:

+	0	1	2	3	4	5		×	0	1	2	3	4	5
0	0	1	2	3	4	5		0	0	0	0	0	0	0
1	1	2	3	4	5	0		1	0	1	2	3	4	5
2	2	3	4	5	0	1		2	0	2	4	0	2	4
3	3	4	5	0	1	2		3	0	3	0	3	0	3
4	4	5	0	1	2	3		4	0	4	2	0	4	2
5	5	0	1	2	3	4		5	0	5	4	3	2	1

As the $+$ table shows, $I/\langle 6 \rangle$ forms a commutative group under addition. (Notice that $-2 \equiv 4, -5 \equiv 1$, and so on.) Associativity of multiplication in $I/\langle 6 \rangle$ can be shown to follow from the same property among the integers. The two distributive properties are also "inherited"; for example,

$$5 \times (3 + 4) \equiv 5 \times 1 \equiv 5$$

and

$$(5 \times 3) + (5 \times 4) \equiv 3 + 2 \equiv 5.$$

Also,

$$(2 + 4) \times 3 \equiv 0 \times 3 \equiv 0$$

and

$$(2 \times 3) + (4 \times 3) \equiv 0 + 0 \equiv 0.$$

The integers modulo 6 form a commutative ring with unit element.

However, $I/\langle 6 \rangle$ is not a field, for some of its non-zero elements do not have multiplicative inverses. Notice in the multiplication table that there is no "1" in the 2 row or in the 2 column. This means that there is no inverse element that multiplies 2 to give 1. Similarly, the table shows that 3 has no inverse and that 4 has no inverse. Notice that 2 and 3 are divisors of $6 \equiv 0$ and that 2, 3, and 4 are divisors of $12 \equiv 0$. An indirect proof that these elements do not have inverses can be based on the fact that they are divisors of zero modulo 6. For instance, suppose 2 had a multiplicative inverse b for which

$$2 \times b \equiv 1.$$

Multiply by 3:

$$3 \times (2 \times b) \equiv 3$$

Reassociate:

$$(3 \times 2) \times b \equiv 3$$

$$0 \times b \equiv 3$$

$$0 \equiv 3$$

Since 3×2 is congruent to zero modulo 6, 2 cannot have an inverse modulo 6. Then the non-zero elements of $I/\langle 6 \rangle$ do not form a group under multiplication, so $I/\langle 6 \rangle$ is not a field.

Exercise 6–74. Find the products 0×2, 1×2, 2×2, $3 \times 2, \dots,$ 9×2 modulo 10, and show that none of the products is 1 (mod 10). Conclude that $I/\langle 10 \rangle$ is not a field.

Exercise 6–75. Continuing the previous exercise, find another integer in $I/\langle 10 \rangle$ that has no inverse.

Exercise 6–76. Find another n besides 6 and 10 for which $I/\langle n \rangle$ is not a field. Show that there is a non-zero element in $I/\langle n \rangle$ for your choice of n that has no multiplicative inverse modulo n.

The integers modulo 5, $I/\langle 5 \rangle$, have arithmetic tables

EXAMPLE

+	0	1	2	3	4
0	0	1	2	3	4
1	1	2	3	4	0
2	2	3	4	0	1
3	3	4	0	1	2
4	4	0	1	2	3

×	0	1	2	3	4
0	0	0	0	0	0
1	0	1	2	3	4
2	0	2	4	1	3
3	0	3	1	4	2
4	0	4	3	2	1

It can be shown that $I/\langle 5 \rangle$ is a ring. Also, the nonzero elements 1, 2, 3, and 4 form a commutative group under multiplication: the group identity is 1; the inverses are $1^{-1} \equiv 1, 2^{-1} \equiv 3, 3^{-1} \equiv 2, 4^{-1} \equiv 4$. It can be shown that $I/\langle 5 \rangle$ is a field.

Exercise 6–77. Form the addition and multiplication tables for $I/\langle 3 \rangle$. Show that the non-zero elements form a commutative group under multiplication. It can be shown that $I/\langle 3 \rangle$ is a field.

Exercise 6–78. Form the addition and multiplication tables for $I/\langle 2 \rangle$ and show that the integers modulo 2 form a field. Notice that it exactly meets the field requirement of "at least two elements."

Exercise 6–79. If n is not a product of two integers both greater than 1, then $I/\langle n \rangle$ is a field (for instance, $I/\langle 5 \rangle$, $I/\langle 7 \rangle$, $I/\langle 23 \rangle$). Show that if n is a product

$$n = km, \text{ with } k > 1 \text{ and } m > 1,$$

then $I/\langle n \rangle$ is not a field.

Some important fields you have used are:

1. the field of rational numbers

2. the field of real numbers

3. the field of complex numbers $s + ti$, where s and t are real numbers. and $i^2 = -1$.

Groups, rings, and fields are examples of "abstract algebras," so to learn more about them, consult books on that topic, for instance, *Abstract Algebra and Solution by Radicals*, by John E. and Margaret W. Maxfield, Chapters 1 and 2 (W. B. Saunders Company, 1971).

REVIEW OF CHAPTER 6

1. Define a group. Give an example of a group, tell what the group operation is, and verify one at a time that the group properties hold.

2. Is the Klein four-group commutative?

3. Define a subgroup. The system $\langle 6 \rangle$ of integral multiples of 6 is a group under addition. Show that it is a subgroup of $\langle 2 \rangle$.

4. What role do negative integers play with respect to integer addition? What role do fractions play with respect to multiplication?

5. Write two different permutations on the set $S = \{1, 2, 3, 4\}$ and form their composition $P \circ Q$. Also form $Q \circ P$. Calculate a composition table for these four permutations. Is each composition one of these original four? Is each composition a permutation on the set S? Do the four permutations form a group?

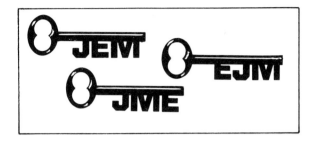

CHAPTER 6 AS A KEY TO MATHEMATICS

By now it seems natural to use the plural arithmetics, for different arithmetics are appropriate to different needs. In this chapter you have been introduced to the classification of arithmetics in general, according to what properties they have. An analogy can be made to the classification of medicines in a pharmacy. Different medicines are developed to treat different disorders. Then the medicines are classified according to an analysis of their similarities and differences, as analgesic, anesthetic, tranquilizing, narcotic, and so on. The classification is helpful in selecting the right medicine for a specific purpose.

Unless you become an algebraist, you will not need to sort arithmetics into groups, semigroups, rings, skew fields, fields, integral domains, lattices, trees, mobs, cliques, coalitions, and so forth. (Not for mathematicians an "Unshapen Land . . . where nothing is in its right place, and nothing has a name"!) However, you already have an important key to mathematics now that you realize there are different arithmetics that can be compared on the basis of their respective properties and applied to the problems they fit best.

CHAPTER 7

A certain tract or parcel of land in the Town of Pittsburg, County of Coos, and State of New Hampshire, being Three Fourths ($\frac{3}{4}$) of Lot Number Two (2) in Range Number One (1) of Lots in said Town of Pittsburg, being the north half of the southwest quarter of said Lot containing approximately One Hundred and Fifty-six acres (156); also approximately Six (6) acres off from the South side of Lot Number One (1) in Range Two (2) of Lots in said Pittsburg.

From a land deed

Here we study a practical subject, as satisfying to one who wants to know what techniques are good for as it is to one whose appreciation is on aesthetic grounds. The simple idea of analytic geometry permeates mathematics subtly, helping us visualize geometrically ideas that would otherwise be expressed only in algebraic equations.

ANALYTIC GEOMETRY

Arithmetic of Location

POINTS

You have advertised a used desk for sale. Someone telephones in response to your classified advertisement and needs directions to your house to see the desk. What kind of directions will help most? Certainly, they should be reasonably short, easy to understand, correct, and not misleading.

Almost at once you need to know where the caller plans to start from, or at least to find a common point of reference that both parties know. Then you must adapt the directions to your caller's experience: Does he know the street names and numbering systems? Route numbers? Geographical features like streams and hills? Manmade landmarks? Compass directions? Has he a measure of distance, a car odometer, for instance?

Although several people independently adopted similiar systems, **cartesian graphs** are named for René Descartes (1596–1650), who developed them. To give the location of any point on a plane, first choose a point in the plane and call it the **origin *O*.** This is like finding a starting point for the desk buyer. Then, through the origin draw a horizontal line called the ***x*-axis** and a vertical line called the ***y*-axis.** This is like telling the desk buyer which streets run east-and-west and which run north-and-south. Next mark the right-hand direction as the **positive** direction on the *x*-axis and the upward direction as positive on the *y*-axis. This is like telling the desk buyer which direction is toward your house on each set of parallel streets. You can use an arrow on each axis to show the positive direction, or you can simply follow the convention that the direction bearing the label "*x*" or "*y*" of the axis is the positive direction. Finally, you need a **unit of measurement**, like the $\frac{1}{10}$ mile or 1 city block you might use in directing the desk buyer. In some cases you may need different units for the two axes, but here suppose the same unit applies. Now you have a picture like the one in Figure 7–1.

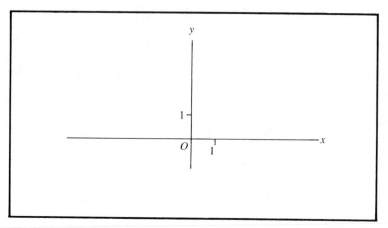

FIGURE 7–1. ELEMENTS OF A CARTESIAN GRAPH.

Refer now to Figure 7–2. To locate the point labeled *P* relative to this axis system, we give it **coordinates** (3, 2). The first coordinate, 3, tells how far to the right of the origin the point *P* is. The second coordinate, 2, tells

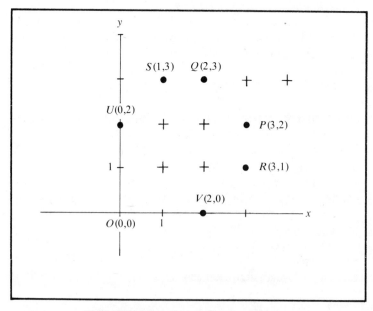

FIGURE 7–2. COORDINATES OF POINTS.

how far above the origin the point P is. The notation $P(3, 2)$ supplies the coordinate of the point along with its name, P.

We follow the standard convention of giving the x coordinate first. Then everyone who knows this code can use it to minimize writing. If you write the 2 before the 3, you have the coordinates of the point $Q(2, 3)$ in Figure 7–2. Note the locations of $R(3, 1)$, $S(1, 3)$, the origin $O(0, 0)$, and $U(0, 2)$ and $V(2, 0)$ on the axes.

Exercise 7–1. Sketch the essentials of a graph, as in Figure 7–1. Graph paper is not required, nor need you do a careful drafting job. A sketch that shows clearly where the following points are is good enough. Plot and label the points

$$M(1, 2) \qquad P(4, 5)$$

$$N(5, 4) \qquad Q(2, 1)$$

Sketch and identify the figure $MQNP$.

Exercise 7–2. Locate and label on a sketched graph the points $S(0, 2)$, $T(2, 0)$, $U(4, 2)$. Locate and give coordinates for a point V that will make the figure $STUV$ a square.

So far, all points have been in one quarter, or quadrant, of the graph. This is because none of the coordinates happened to be negative. Figure 7–3 shows points in various quadrants. The point P has coordinates

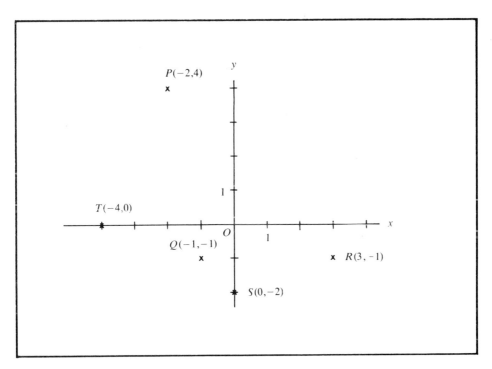

FIGURE 7–3. Points with negative coordinates.

(−2, 4). The first coordinate is supposed to tell how far to the right of the origin P is. Then P is −2 units to the right. What does the negative mean here? We return to the root meaning of a negative 2 as the number that gives zero when +2 is added to it. Where is a point located if the point +2 units to its right is zero units from the y-axis? It is 2 units to the *left* of the y-axis. Note the locations of $Q(-1, -1)$, $R(3, -1)$, $S(0, -2)$, and $T(-4, 0)$.

Exercise 7–3. On a cartesian graph, locate and label the points

$$P(3, 4) \qquad R(-3, -4)$$

$$Q(-3, 4) \qquad S(3, -4)$$

Exercise 7–4. Find the coordinates for each of the 20 points named in Figure 7–4 (D through W, including the origin O). Save your work for later reference.

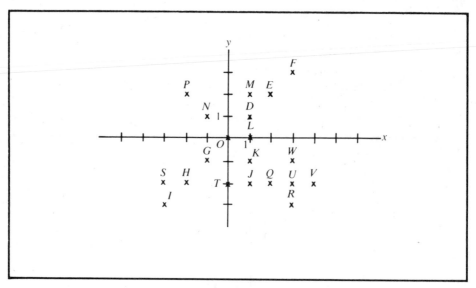

FIGURE 7–4. LOCATION OF POINTS.

Exercise 7–5. Suppose you live in a planned city with the origin O at City Hall, the city block a reliable unit, east the positive x-direction, north the positive y-direction. Sketch a map-graph showing the school 4 blocks east and 5 blocks south of City Hall, the Meeting House 10 blocks west and 1 block north of City Hall, the jail 2 blocks west and 2 blocks south of City Hall, and the library 3 blocks due west of City Hall on the same east–west Avenue. Save your map for Exercise 7–7.

Now you know how to label any point in the plane that happens to fall on the lattice of graph lines drawn at unit intervals. To label a point that is between two unit lines on a graph, you need decimal coordinates. Figure 7–5 shows the points $P(1, \frac{1}{2})$, $Q(-2\frac{1}{2}, 1\frac{2}{3})$, $R(4.6, 0)$, and $S(3.5, -2.1)$.

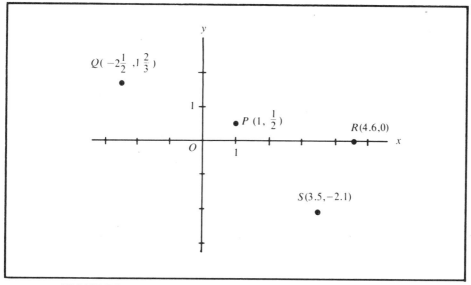

FIGURE 7–5. POINTS WITH NONINTEGRAL COORDINATES.

Exercise 7–6. Plot and label the following points on a cartesian graph. Commercial graph paper would be helpful here, but you can make an adequate substitute with a little extra work.

$$P(3.4, 1.5) \qquad Q(2.3, -3.5) \qquad R(-2.4, -2.4)$$
$$S(-1.0, 2.7) \qquad T(-4.1, -3.9)$$

Exercise 7–7. Return to the map of Exercise 7–5. The floating crap game just docked embarrassingly near City Hall, in an alley halfway between the avenues 2 and 3 blocks, respectively, north of City Hall, and $1\frac{1}{4}$ blocks west of City Hall. Locate the game on your map and label it with its coordinates in the conventional order.

You now have all the information you need to locate any point in a plane. As you have discovered while working with rulers and graph paper, there are practical limitations on the accuracy of locating a point like $(3.127, -2.146)$, for instance, but at least theoretically every point in the plane has an x-coordinate and a y-coordinate that pinpoint it unambiguously. Conversely, any two signed decimal numbers given in a definite order (x, y) determine exactly one point in the plane.

During World War II there was a scientist who needed to receive the results of laboratory experiments while he was on trips away from the laboratory. He worked out a code with his laboratory assistants, based on the importance of the origin and axes and unit on a cartesian graph. Before leaving for a trip, he agreed with his assistants on the location on graph paper of these essentials. Then when the results were ready, his assistants entered them as points, lines, curves, and so on without drawing the axes. The information in this form was mailed unclassified with no special precautions, since without the reference system the results were completely incomprehensible to anyone who intercepted them.

The cartesian correspondence between points and number pairs is amazingly fruitful. With it you can describe geometrical figures in the plane by algebraic formulas or equations, so that routine algebraic manipulation, suitable for computer programs, can be used to solve the more intuitive problems of geometry.

LINES

Horizontal Lines, Vertical Lines

Refer to Figure 7–4. Notice that the points S, H, T, J, Q, U, and V all lie on a horizontal line. Note the coordinates for these points (Exercise 7–4). What do all the coordinates have in common?

Exercise 7–8. Sketch a graph and plot these points having y-coordinate 3: $A(1, 3)$, $B(-4, 3)$, $C(-2\frac{1}{2}, 3)$, $D(3, 3)$, and $E(0, 3)$.

Exercise 7–9. Find the coordinates of the points lettered on line KK' in Figure 7–6. What do they have in common?

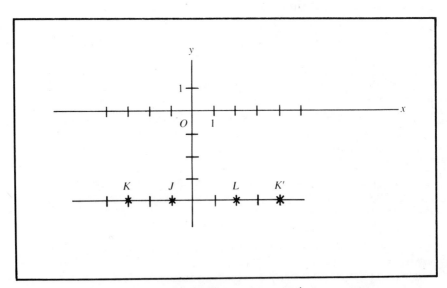

FIGURE 7–6. LINE KK'.

You can give a formula for the horizontal line 5 units above the origin:

$$y = 5.$$

Any point (x, y) with coordinates that satisfy the equation lies on the line: $(2, 5)$, $(-1, 5)$, $(3.33, 5)$, and so on. The formula makes no restriction on the first, or x-coordinate, because there is a position on the horizontal line for every x value.

The formula for a horizontal line k units above the origin is

$$y = k.$$

Here k can be positive, negative, zero, integral, or fractional.

Where are all points of the plane whose coordinates satisfy the for- **EXAMPLE** mula

$$y = -2\tfrac{1}{2}?$$

They are points such as $(3, -2\tfrac{1}{2})$, $(-2, -2\tfrac{1}{2})$, $(0, -2\tfrac{1}{2})$, and so on, having y-coordinate $-2\tfrac{1}{2}$. They all lie on a horizontal line $2\tfrac{1}{2}$ units below the origin.

Where are the points whose coordinates meet the restriction

$$y = 0?$$

Some examples are $(0, 0)$, $(1, 0)$, $(-5, 0)$, and $(10, 0)$. They all lie on the x-axis.

Exercise 7–10. Sketch on a graph 5 points whose coordinates satisfy the equation

$$y = 7,$$

and draw the line $y = 7$ (at least as much of it as will fit on your sketch!).

Exercise 7–11. A weaver is dressing his loom with warp threads as shown in Figure 7–7. The bottom black stripe is bordered by the horizontal lines $y = -4$ and $y = -3\tfrac{1}{2}$. Give the equations of the two lines that border each of the other 8 stripes. Save your work to use in a later exercise. Do you see how cartesian geometry could be used to program an automated loom?

The vertical line 3 units to the right of the origin has the formula or equation

$$x = 3,$$

because all the points on the line have the x-coordinate 3: $(3, 4)$, $(3, -1\tfrac{1}{2})$, $(3, 2)$, $(3, 0)$, for instance. The equation for the vertical line h units to the right of the origin is

$$x = h.$$

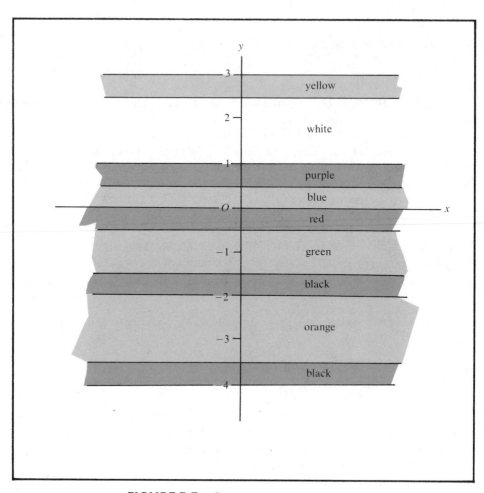

FIGURE 7–7. COLOR PATTERN FOR DRESSING A LOOM.

The equation $x = -2$ describes the vertical line 2 units to the left of **EXAMPLE** the origin, for without restricting the y-coordinate, it requires the x-coordinate of each point to be -2.

The equation of the y-axis is $x = 0$.

Exercise 7–12. Label 5 points on the y-axis and note that each has x-coordinate zero.

Exercise 7–13. Refer to Figure 7–7 and Exercise 7–11. The weaver has dressed his loom with warp threads as shown. Now he weaves across them with vertical woof threads in stripes bordered as follows:

$$\text{black from } x = -5 \quad \text{to} \quad x = -4\tfrac{1}{2}$$
$$\text{blue from } x = -4\tfrac{1}{2} \quad \text{to} \quad x = -3\tfrac{1}{2}$$
$$\text{white from } x = -3\tfrac{1}{2} \quad \text{to} \quad x = -3$$
$$\text{green from } x = -3 \quad \text{to} \quad x = -2$$
$$\text{pink from } x = -2 \quad \text{to} \quad x = -1$$
$$\text{red from } x = -1 \quad \text{to} \quad x = 2.$$

Sketch the woven piece, coloring it, or at least labeling the color of each stripe. What are its total dimensions? Give the 4 lines (by their equations) that border an all-green square. Give the bordering lines for both blue-white rectangles of plaid.

Sometimes we write $x = -2$ in the form $x + 2 = 0$, $y = 7$ in the form $y - 7 = 0$, and so on. Both forms describe the same line.

The points $(5, -3)$, $(10, -3)$, $(0, -3)$, $(1, -3)$ lie on the horizontal **EXAMPLE** line $y = -3$. The coordinates in each pair satisfy the equation

$$y + 3 = 0.$$

Exercise 7–14. Sketch on a graph the lines

$$x - 1 = 0 \qquad y + 2 = 0$$
$$x + 3 = 0 \qquad y - 6.5 = 0$$

You are designing patterns for a card section at a football game. You **EXAMPLE** assign x and y coordinates to each seat in the cheering section, x giving the distance from the 50-yard line, and y giving the level above or below the stadium entrance. Figure 7–8 shows the word "FELT" printed on a

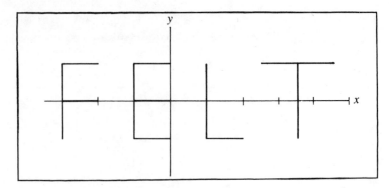

FIGURE 7-8. LETTERS ON A GRAPH.

graph. The letter E could be described as follows:

$$E \begin{cases} x + 1 = 0 & \text{from} \quad y = -1 \quad \text{to} \quad y = +1 \\ y + 1 = 0 & \text{from} \quad x = -1 \quad \text{to} \quad x = 0 \\ \quad y = 0 & \text{from} \quad x = -1 \quad \text{to} \quad x = 0 \\ y - 1 = 0 & \text{from} \quad x = -1 \quad \text{to} \quad x = 0 \end{cases}$$

(All four line segments taken together.)

Exercise 7–15. Following the Example, give equation descriptions for the letters F, L, and T in Figure 7–8.

Exercise 7–16. What letter has the following description?

$$\begin{cases} x + 7 = 0 & \text{from} \quad y = -5 \quad \text{to} \quad y = -1 \\ x + 3 = 0 & \text{from} \quad y = -5 \quad \text{to} \quad y = -1 \\ y + 3 = 0 & \text{from} \quad x = -7 \quad \text{to} \quad x = -3 \end{cases}$$

Lines Through the Origin

In Figure 7–4 the points I, H, G, O, D, E, and F lie on a diagonal line through the origin. Examine their coordinates (Exercise 7–4) to see what they have in common. Does the point $(7, 7)$ lie on the same line? Does $(7, -7)$? Does $(-7, -7)$? Does $(-7, 7)$? Since each point on the line has its two coordinates equal, the equation of the line is

$$y = x.$$

This diagonal line is shown in Figure 7–9.

Exercise 7–17. Plot and give coordinates for 4 points on the line $y = x$ shown in Figure 7–9. Verify that the coordinates of the points satisfy the equation $x - y = 0$. Plot $(1, -1)$, $(-2, 2)$, $(3, 4)$, and $(-6, -7)$, showing that they do not lie on the line $y = x$. Show that their coordinates do not satisfy the equation $x - y = 0$.

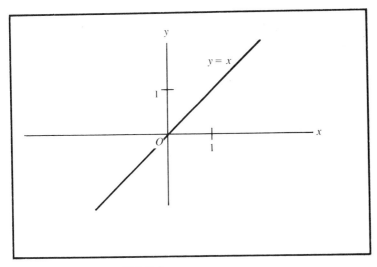

FIGURE 7–9. THE LINE $y = x$.

Exercise 7–18. Plot the points $(1, -1)$, $(2, -2)$, $(3, -3)$, $(-1, 1)$, $(-2, 2)$, and $(-3, 3)$. Show that the coordinates in each pair satisfy the equation

$$x + y = 0.$$

Then draw the line through the 6 points and label it $x + y = 0$. The origin O lies on the line. Verify that the coordinates of the origin satisfy the equation $x + y = 0$.

Exercise 7–19. An industrial company has a matching program: for each dollar an employee contributes to his college the company contributes a dollar also. Make a table showing the company's contribution for an employee contribution of $0, $1, $5, $10, $15, $25, $100, $500, and $1000. (Leave a blank column in your table to use in Exercise 7–20.) Now graph these 9 points, using the x-axis for e, the employee's contribution, and the y-axis for c, the company's contribution. (In sketching this graph, notice that you will need no negative coordinates, and that the scale must be chosen to cover from $0 to $1000.) Write the formula for c in terms of e. Draw the line of points whose coordinates satisfy the formula.

Exercise 7–20. Continuing the previous exercise, let r stand for the amount the college receives from the employee and the company together.

In the column you left blank, table the r value for each e value; for instance, opposite $e = \$50$, the r value should be $\$100$, for the college receives $r = e + c = \$50 + \$50$. Now make a cartesian graph using the x-axis for e and the y-axis for r. Plot the 9 points from your table. Connect them with a line. Notice that points on the line have coordinates (x, y) for which $y = 2x$.

In Figure 7–10 two lines through the origin are shown. The line labeled $y = \frac{1}{2}x$ is made up of points $(2, 1)$, $(3, 1\frac{1}{2})$, $(-2, -1)$, $(3.14, 1.57)$, and so on, with the second, or y, coordinate equal to one-half the x coordinate. The line labeled $y = -3x$ is made up of points $(1, -3)$, $(2, -6)$, $(-\frac{1}{3}, 1)$, $(-2, 6)$, and so on, with y-coordinate -3 times the x-coordinate.

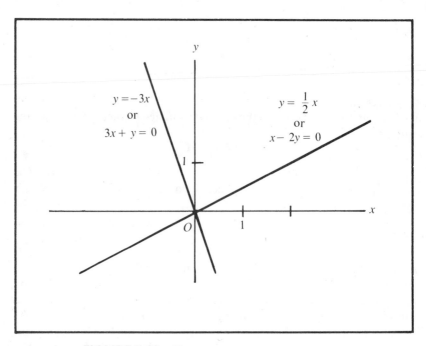

FIGURE 7–10. Two lines through the origin.

Slope m; Parameter; Family

Suppose a line through the origin passes also through the point $M(1, m)$, as shown in Figure 7–11. Let $P(x, y)$ be another point on the

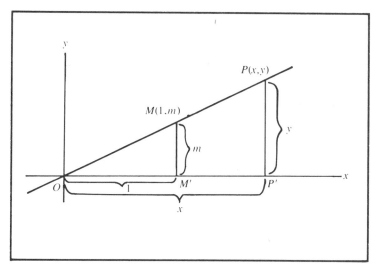

FIGURE 7-11. THE SLOPE RATIO y/x.

The ratio y/x is constant for all points on a line.

line. Then the triangles MOM', with M' the point $(1, 0)$, and POP', with P' the point $(x, 0)$, are "similar" triangles; that is, their respective angles are equal, although the triangles are of different sizes. We use here the fact from plane geometry that the respective sides of similar triangles are proportional:

$$\frac{y}{x} = \frac{m}{1},$$

which can be rewritten in the form

$$y = mx.$$

Every point, including the origin, having the y-coordinate m times the x coordinate lies on the line, and its coordinates satisfy the equation $y = mx$, which is called the equation of the line.

If m is positive, then the larger x is, the larger y is, so that the line slopes upward to the right. If m is negative, then the larger x is the smaller y is, so the line slopes downward to the right. If m is zero, then y has constant value zero, and the line is the x-axis, $y = 0$.

Then as m takes on different values, the equation $y = mx$ represents a **family** of different lines through the origin. The only line left out of the family is the y-axis, $x = 0$. The **parameter** m, whose value determines which line of the family is meant, is called the **slope** of the line $y = mx$.

EXAMPLE Graph $y = \frac{2}{3}x$. This equation represents one of the family of lines $y = mx$ through the origin, the one with slope $m = \frac{2}{3}$. Start from the origin. For each unit increase in x, y increases $\frac{2}{3}$ unit. The points $(1, \frac{2}{3})$, $(3, 2)$, $(6, 4)$, $(-1, -\frac{2}{3})$, $(-3, -2)$ lie on the line, which is shown in Figure 7-12.

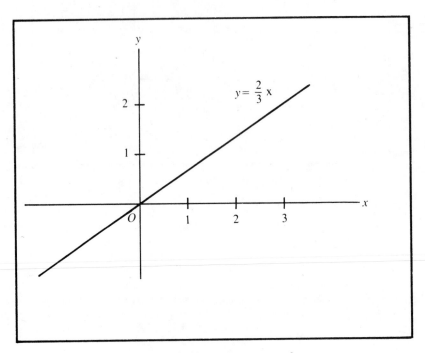

FIGURE 7-12. THE LINE $y = \frac{2}{3}x$.

Exercise 7-21. Draw on one cartesian graph the lines through the origin corresponding to the following equations. Use dotted lines or colored pencil for those that have a negative slope. Find a point other than the origin on each line, give its coordinates, and verify that the coordinates satisfy the equation.

$$y = \tfrac{1}{3}x \qquad 2x + y = 0 \text{ (Write in } y = mx \text{ form.)}$$
$$y = -2x \qquad y = 4.5x$$
$$y = -9x \qquad -3x + 2y = 0 \text{ (Write as } y = \tfrac{3}{2}x.\text{)}$$

Exercise 7-22. A line can be labeled PQ in terms of two points on it, P and Q. In Figure 7-13 find the slopes of the lines OA, OB, OC, OD, OE, OF, and OG.

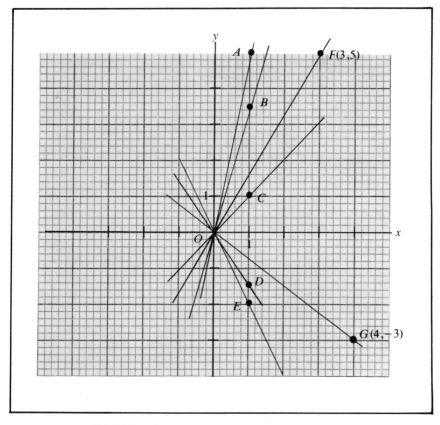

FIGURE 7–13. LINES OF THE FAMILY $y = mx$.

Exercise 7–23. In Figure 7–13 find a point A' on the line OA the same distance from O as A, but in the opposite direction. You will find that the coordinates of A' are $(-1, -5)$. Find the coordinates of points B', C', D', E', F', and G', similarly defined.

Exercise 7–24. Let x stand for yards and y stand for feet. Each additional 3 feet contribute 1 additional yard. Graph the line $y = 3x$ for non-negative values. Verify for 5 values that your graph converts lengths given in feet into yards. Can it be used to convert yards into feet?

Exercise 7–25. The formula $d = rt$ gives the distance traveled as the rate r times the time t. For instance, if you travel at $r = 60$ miles an hour for $t = 3$ hours, you travel a distance of $d = 3 \times 60 = 180$ miles. Using the x-axis for t values and the y-axis for d values, graph lines of the family $d = rt$ for 3 different values of the parameter r. Suppose you want to travel 280 miles in $3\frac{1}{2}$ hours. Find what rate of travel you need, and graph the corresponding line.

Exercise 7–26. You are a hack writer, earning $1\frac{1}{2}$¢ a word for pulp fiction. Draw a chart that will tell you your earnings for each 1000 words (use 1000 words as the x "unit").

Lines with y-Intercept b

Figure 7–14 shows the line $y = \frac{1}{2}x$, and 6 points on this line, $(0, 0)$, $(1, \frac{1}{2})$, $(2, 1)$, $(-1, -\frac{1}{2})$, $(-2, -1)$, and $(4, 2)$, indicated by triangles.

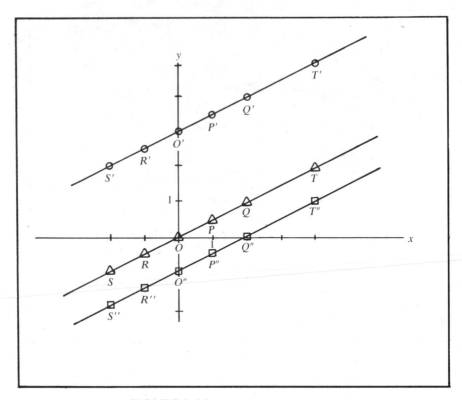

FIGURE 7–14. LINES PARALLEL TO $y = \frac{1}{2}x$.

The points indicated by circles are located 3 units above the corresponding points on $y = \frac{1}{2}x$; that is, each y value for a circle-point is 3 units more than the y value for the triangle-point.

The point S has coordinates $(-2, \frac{1}{2}(-2))$, or $(-2, -1)$. Then S' has coordinates $(-2, -1 + 3)$, or $(-2, 2)$. The coordinates of O' are $(0, 0 + 3)$, or $(0, 3)$.

Exercise 7–27. Find the coordinates of R', P', Q', and T'.

Since each point on the line $y = \frac{1}{2}x$ has y-coordinate $\frac{1}{2}x$, the corresponding circle-point 3 units above has y coordinate $y = \frac{1}{2}x + 3$. In fact, all the points whose coordinates satisfy the equation

$$y = \frac{1}{2}x + 3$$

form a line 3 units above the line through the origin

$$y = \frac{1}{2}x,$$

in the sense that for each x the y values are 3 units apart.

The points O'', P'', Q'', and so on, indicated by squares, are located one unit below the corresponding triangle-points on the line $y = \frac{1}{2}x$. All the points whose coordinates satisfy the equation

$$y = \frac{1}{2}x - 1$$

form a line one unit below the line through the origin

$$y = \frac{1}{2}x.$$

Exercise 7–28. Find the coordinates of the 6 square-points, and show that they satisfy the equation

$$y = \frac{1}{2}x - 1.$$

To write the equation of the line b units above $y = \frac{1}{2}x$ at each x value, add b to each y coordinate, getting

$$y = \frac{1}{2}x + b.$$

If b is negative, the line is below the origin. If b is zero, the line passes through the origin and is the line

$$y = \frac{1}{2}x.$$

The origin $(0, 0)$ lies on the line $y = \frac{1}{2}x$. The corresponding point on the y-axis of the line $y = \frac{1}{2}x + b$ is the point $(0, b)$; and b is called the **y-intercept** of the line $y = \frac{1}{2}x + b$.

Figure 7–15 shows the line made up of points b units above the corresponding points of the line through the origin $y = mx$. At $x = 0$, the

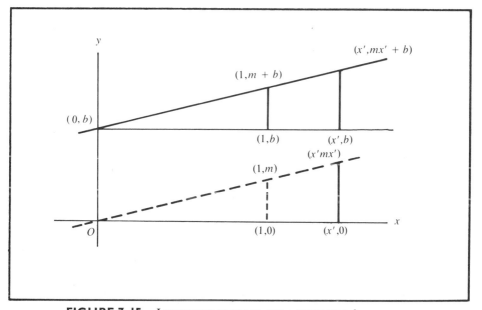

FIGURE 7–15. LINES WITH SLOPE m AND y-INTERCEPT b.

The line $y = mx + b$ is parallel to the line $y = mx$ and b units above it, cutting the y-axis at $y = b$.

point b units above the line $y = mx$ is $(0, b)$. At $x = 1$, the point on the line $y = mx$ is $(1, m)$, so the point b units above it is $(1, m + b)$. At $x = x'$, the point on the line $y = mx$ is (x', mx'), so the point b units above it is $(x', mx' + b)$, whose coordinates satisfy the equation

$$y = mx + b.$$

The **slope** of the line $y = mx + b$ is defined to be the same as the slope of $y = mx$, or m. Since the two lines remain the same vertical distance apart, they do not intersect, and so are "**parallel**" lines, as in geometry. The parameter b is called the **y-intercept** of the line $y = mx + b$.

EXAMPLE   Graph lines of the family

$$y = -3x + b,$$

for 5 different values of the parameter b.

First, graph the line $y = -3x$ for which the parameter b takes the value zero (Figure 7–16).

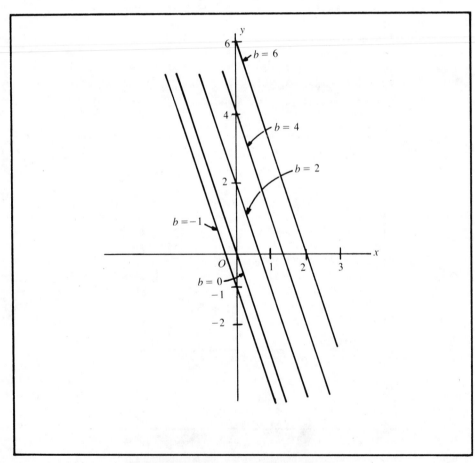

FIGURE 7–16. LINES OF THE FAMILY $y = -3x + b$.

To graph the line of the family for which $b = -1$, plot points 1 unit below the line $y = -3x$, connecting them by the line parallel to $y = -3x$ and one unit below. The line of the family for which $b = 6$ is parallel to $y = -3x$ and 6 units above it.

Exercise 7-29. Graph the line $y = 2x$. Graph $y = 2x + 4$ and $y = 2x - 2$.

Exercise 7-30. Referring to Figure 7–17, write the equations of OB and the two lines parallel to it. Write the equation of their family, and give the corresponding three values of the parameter b.

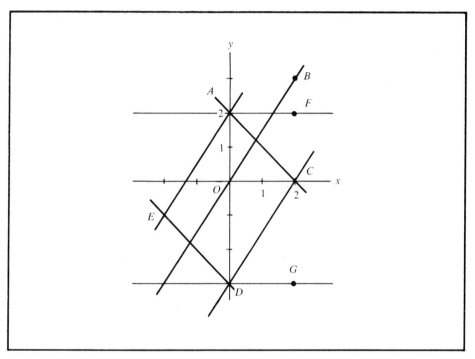

FIGURE 7–17. LINES OF SEVERAL FAMILIES.

In Figure 7–17 the lines DE, DC, and DG all have y-intercept $b = -3$ [at $D(0, -3)$], but their slopes differ. The lines can be described by equations: **EXAMPLE**

$$DE: y = -x - 3$$
$$DC: y = \tfrac{3}{2}x - 3$$
$$DG: y = 0x - 3, \quad \text{or} \quad y = -3.$$

Their family equation is $y = mx - 3$, with the slope m as parameter.

Exercise 7-31. Write the equations of AE, AC, and AF, and their family equation (Figure 7–17).

Every **line** in a plane has a **corresponding linear equation** of the form

$$Ax + By + C = 0, \quad A \text{ and } B \text{ not both } 0,$$

and conversely, every linear equation can be represented on a cartesian graph as a straight line. To substantiate the first part of this statement, suppose L is a straight line in a plane having a cartesian coordinate system. If L is vertical, it has the equation

$$x = h,$$

which is a linear equation with $A = 1$ $B = 0$, and $C = -h$. If L is not vertical, then it has a slope m and a y-intercept b, so that its equation is

$$y = mx + b.$$

This equation is linear with $A = m$, $B = -1$, and $C = b$. To prove the second, or converse part of the statement, consider the linear equation

$$Ax + By + C = 0, A \text{ and } B \text{ not both } 0.$$

Suppose B is zero. Then A is not zero. Multiplying both sides of the equation by $1/A$, we have

$$\left(\frac{1}{A}\right)Ax + \left(\frac{1}{A}\right)0y + \left(\frac{1}{A}\right)C = 0,$$

or

$$x + C/A = 0,$$

or

$$x = -C/A.$$

This equation can be represented by the vertical line $-C/A$ units to the right of the y-axis.

Now suppose B is not zero. Multiplying the linear equation by $1/B$, we have

$$\left(\frac{1}{B}\right)Ax + \left(\frac{1}{B}\right)By + \left(\frac{1}{B}\right)C = 0,$$

or

$$\left(\frac{A}{B}\right)x + y + \frac{C}{B} = 0,$$

or

$$y = -\left(\frac{A}{B}\right)x - \frac{C}{B}.$$

This equation can be represented by a line with slope $-(A/B)$ and y-intercept $-C/B$.

Exercise 7–32. Check which of these equations represent lines that slant upward to the right. Put a letter A beside those that represent lines above the origin (positive y-intercept). Which slant neither up nor down? Which pass through the origin?

1. $y = 121x - 10$ 6. $y = 3x + 7$

2. $y = -3x - 1$ 7. $y = -2x + 5$

3. $y = 0.01x + 0.01$ 8. $y = 6$

4. $y = x - 2$ 9. $y = -100x + 100$

5. $y = 2x$ 10. $y = 0$

Figure 7–18 shows the letters "$M\ A\ T\ H$" drawn on a cartesian graph. **EXAMPLE**

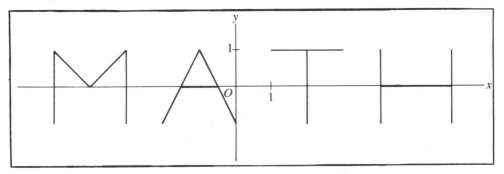

FIGURE 7–18.   Letters on a graph.

The letter M could be given by equations as follows:

$$M \begin{cases} x = -5, \text{ from } y = -1 \text{ to } y = +1 \text{ (or } -1 \leq y \leq +1) \\ y = -x - 4, \text{ with } x \text{ in the range } -5 \leq x \leq -4 \\ y = x + 4, \text{ with } -4 \leq x \leq -3 \\ x = -3, -1 \leq y \leq +1 \end{cases}$$

The letter H could be given by

$$H \begin{cases} x = 4, -1 \leq y \leq +1 \\ y = 0, \quad 4 \leq x \leq 6 \\ x = 6, -1 \leq y \leq +1 \end{cases}$$

Exercise 7–33.   Write the letter A in Figure 7–18 in the form of equations.

Exercise 7–34.   Draw the figure represented by these conditions, all line segments drawn on a common graph:

(a) $y = 2x + 3, -1 \leq x \leq -\frac{1}{2}$ (f) $x = 2, -\frac{1}{2} \leq y \leq 1$

(b) $y = -2x + 1, -\frac{1}{2} \leq x \leq 0$ (g) $y = x - 2\frac{1}{2}, 2 \leq x \leq 3$

(c) $y = -2x - 1, -1 \leq x \leq -\frac{1}{2}$ (h) $y = -x + 3\frac{1}{2}, 3 \leq x \leq 4$

(d) $y = 2x + 1, -\frac{1}{2} \leq x \leq 0$ (i) $x = 4, -\frac{1}{2} \leq y \leq 1$

(e) $y = 1, 0 \leq x \leq 4\frac{1}{2}$

Intersecting Lines

Each point on the line $y = 2x - 1$ has coordinates that satisfy the equation. This includes points P, Q, and R of Figure 7–19. Each point on

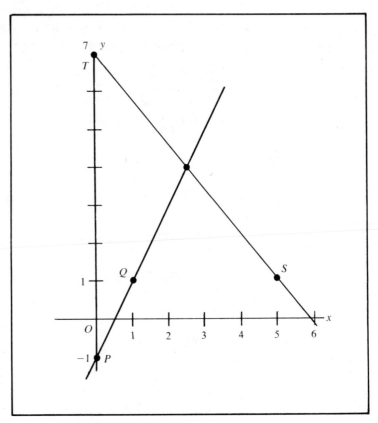

FIGURE 7–19. Two intersecting lines.

the line $y = -\frac{6}{5}x + 7$ has coordinates that satisfy the equation. This includes the points R, S, and T of Figure 7–19. Since the intersection point R lies on both lines, its coordinates satisfy both equations:

$$\begin{cases} \hat{y} = 2\hat{x} - 1 \\ \hat{y} = -\frac{6}{5}\hat{x} + 7. \end{cases}$$

To find what values $\hat{x}$, $\hat{y}$ make both equations hold, we find the simultaneous solution of the equations in this way: Set the two formulas for $\hat{y}$ equal to each other, since R has the common height $\hat{y}$ on both lines.

$$2\hat{x} - 1 = -\tfrac{6}{5}\hat{x} + 7,$$

from which

$$2\hat{x} + \tfrac{6}{5}\hat{x} = +7 + 1,$$

so that

$$3\tfrac{1}{5}\hat{x} = 8$$
$$\tfrac{16}{5}\hat{x} = 8$$
$$\tfrac{2}{5}\hat{x} = 1$$
$$\hat{x} = \tfrac{5}{2}$$

Then $\hat{y} = 2(\tfrac{5}{2}) - 1 = 5 - 1 = 4$.

As a check, show that $R(\tfrac{5}{2}, 4)$ lies on both lines.

$$4 \overset{?}{=} 2(\tfrac{5}{2}) - 1 \qquad\qquad 4 \overset{?}{=} -\tfrac{6}{5}(\tfrac{5}{2}) + 7$$
$$4 = 5 - 1. \quad \text{Yes.} \qquad\qquad 4 \doteq -3 + 7. \quad \text{Yes.}$$

Exercise 7-35. Show that the lines $y = x - 5$ and $y = -2x + 4$ intersect at $P(3, -2)$, by showing that the coordinates of P satisfy both equations. Graph both lines and locate P.

Exercise 7-36. Graph these two lines carefully:

$$\begin{cases} y = -x + 3 \\ y = 3x - 5. \end{cases}$$

Estimate the coordinates of the intersection from your graph. Check to see whether the estimates satisfy both equations.

Exercise 7-37. Find the intersection of these lines by estimating from a graph. Solve the two equations simultaneously to check your estimation.

$$\begin{cases} y = 4x - 2. \\ y = -2x + 4. \end{cases}$$

Exercise 7-38. Solve these equations simultaneously for x and y:

$$\begin{cases} y = -2x + 8 \\ y = 4x - 12 \end{cases}$$

Graph the lines and find the coordinates of their intersection.

Exercise 7-39. Your income from your patented Gizmatic consists of an initial grant for Research and Development of $b = \$100$, plus $m = \$125$ for each Gizmatic you sell. Your expenses consist of an initial outlay of $b' = \$250$ to start your business, plus $m' = \$75$ cost for making each

Gizmatic. Find your break-even point, the number of Gizmatics you must make and sell to have your total expense equal your total income, by graphing the lines $y = mx + b$ and $y = m'x + b'$ and finding their intersection.

Exercise 7-40. There is no simultaneous solution for the equations

$$\begin{cases} 2x - y + 7 = 0 \\ -4x + 2y + 1 = 0. \end{cases}$$

To check this statement algebraically, add the first equation to $\frac{1}{2}$ times the second. Then check geometrically by graphing the two lines represented by the equations, showing that they do not intersect. What is the slope m for each line? What is the y-intercept b?

Two lines that are not parallel intersect at exactly one point. That one point provides the only coordinates, then, that satisfy both linear equations simultaneously. The geometric fact that the two lines have exactly one intersection is reflected in the algebraic fact that the two linear equations have exactly one simultaneous solution.

Suppose two lines are parallel. How is this geometric fact reflected algebraically? Since parallel lines do not have a point of intersection, there is no coordinate pair that can satisfy both linear equations at once. There is no simultaneous solution for the two equations.

Exercise 7-41. Vladimir could jog 20 minutes the first day and improves 2 minutes each day. Enrico could jog 10 minutes the first day and improves 3 minutes each day. In how many days will Enrico be able to jog as long as Vladimir?

Exercise 7-42. Sophie starts from Boston and drives north at 50 miles per hour. Karl also starts north on the same road at the same time Sophie does, but he starts from 40 miles south of Boston. He travels at 52 miles per hour. How long will it take Karl to overtake Sophie? How far north of Boston will they be when they meet?

DISTANCES AND CIRCLES

In Figure 7-20 how far is it from point $P(2, 1)$ to point $Q(10, 1)$? The distance is $10 - 2 = 8$ units. How far is it from P to $R(2, 6)$? The distance is $6 - 1 = 5$ units.

Exercise 7-43. Find the distance between S and T. Find the distance between T and U.

To find the distance between $V(-3, 5)$ and $T(9, 5)$, use the x coordinates as signed numbers. Notice in Figure 7-20 that the distance VT is the

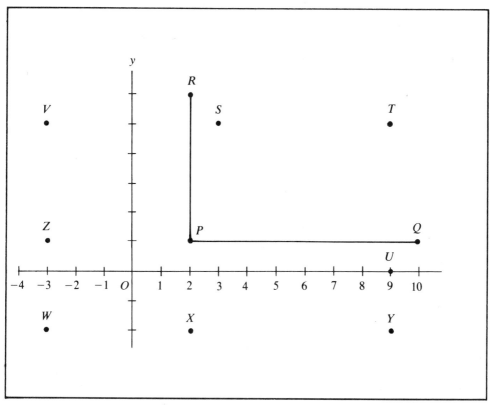

FIGURE 7–20. Distances between points.

9-unit distance from the y-axis to T plus the 3-unit distance from V to the y-axis.

$$9 - (-3) = 9 + 3 = 12$$

The distance between $W(-3, -2)$ and $Z(-3, 1)$ is   **EXAMPLE**

$$1 - (-2) = 1 + 2 = 3,$$

The distance between $Y(9, -2)$ and $T(9, 5)$ is

$$5 - (-2) = 5 + 2 = 7.$$

Exercise 7–44. Find these distances in Figure 7–20 by subtracting signed numbers. Check by counting units in the figure.

ST	WV	XY
VS	XP	YU
ZV	XR	UT

Exercise 7–45. The temperature drops from $15°$ to $-5°$. How many degrees does it drop?

Exercise 7–46. You owed $2, that is, you had savings of -2 dollars.

Then your ship came in, and now you have $+\$3$. How much money did you gain?

The **horizontal distance** between points $C(h, k)$ and $Q(x, k)$ is $x - h$. The **vertical distance** between $Q(x, k)$ and $P(x, y)$ is $y - k$. (See Figure 7–21.) To find the distance between C and P, you can use a famous result from geometry called the **Pythagorean Theorem:** In a right triangle the square of the hypotenuse (longest side) length equals the square of one leg length plus the square of the other leg length.

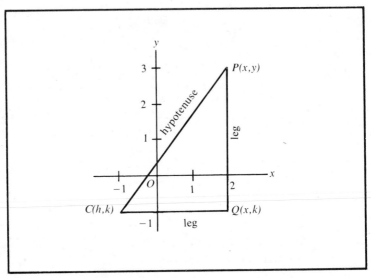

FIGURE 7–21. DIAGONAL DISTANCE.

The triangle CQP of Figure 7–21 is a right triangle, because PQ is vertical and CQ horizontal, so that they form a right angle at Q. The Pythagorean theorem makes this statement about the lengths:

$$(CP)^2 = (CQ)^2 + (QP)^2.$$

By subtraction you have found the horizontal length

$$CQ = x - h$$

and the vertical length

$$QP = y - k.$$

Then

$$(CP)^2 = (x - h)^2 + (y - k)^2.$$

EXAMPLE In Figure 7–21, take the coordinates to be $C(-1, -1)$, $Q(2, -1)$, $P(2, 3)$. Then $CQ = 2 - (-1) = 3$, and $QP = 3 - (-1) = 4$. Then

$$(CP)^2 = 3^2 + 4^2$$
$$= 9 + 16$$
$$= 25.$$

Then CP must be 5, since $5^2 = 25$.

Find the distance ZT in Figure 7–20.   **EXAMPLE**
The coordinates are $Z(-3, 1)$ and $T(9, 5)$. From the Pythagorean theorem,

$$(ZT)^2 = (9 - (-3))^2 + (5 - 1)^2$$
$$= 12^2 + 4^2$$
$$= 144 + 16$$
$$= 160.$$

Then to find the distance ZT, look for a number d that gives 160 when squared:

$$d^2 = 160 = 16 \cdot 10 = 4^2 \cdot 10.$$

Equivalently, you can look for a number d' with $d'^2 = 10$, since then you can use $4d'$ as d:

$$d^2 = (4d')^2 = 16d'^2 = 160.$$

An excellent experimental method for approximating $\sqrt{10}$ is to try squaring numbers until you approximate 10 (Figure 7–22). Another perfectly good method is to look up the answer in a table or chart (Figure 7–23).

$$3^2 = 9, \qquad 4^2 = 16, \quad \text{so} \quad 3 < d' < 4$$

3.2	3.1	3.16	3.17
$\times 3.2$	$\times 3.1$	$\times 3.16$	$\times 3.17$
64	31	1896	2219
$9\ 6$	$9\ 3$	316	317
10.24	9.61	$9\ 48$	$9\ 51$
		9.9856	10.0489

FIGURE 7–22. FINDING $\sqrt{10}$ BY EXPERIMENT.

Since d' is between 3.16 and 3.17, we have $d = 4d'$ is between 12.64 and 12.68, or to the nearest tenth, $12.6 < d < 12.7$.

The **distance formula** for the distance between $C(h, k)$ and $P(x, y)$ is

$$d = CP = \sqrt{(x - h)^2 + (y - k)^2}$$

Exercise 7–47. Find the following lengths in Figure 7–20, locating them between two tenths, as in the last Example.

$$PT$$
$$XU$$
$$OP$$

Exercise 7–48. Graph the following points and show by the distance formula that they are all the same distance from the point $C(2, 3)$:

$$(6, 6) \quad (-1, 7) \quad (6, 0)$$
$$(5, 7) \quad (-2, 6) \quad (-1, -1)$$
$$(2, 8) \quad (-3, 3) \quad (3, 3 + \sqrt{24})$$

As Exercise 7–48 suggests, the distance formula can be used to describe a circle algebraically in terms of a cartesian graph. Since a circle is made up of all the points $P(x, y)$ that have the common distance r (for "radius") from the center $C(h, k)$, the **formula for a circle** is

$$\sqrt{(x - h)^2 + (y - k)^2} = r.$$

This is usually written with both sides squared:

$$(x - h)^2 + (y - k)^2 = r^2.$$

EXAMPLE   The points in Exercise 7–48 lie on the circle with formula

$$(x - 2)^2 + (y - 3)^2 = 5^2.$$

Sometimes the formula appears with the terms squared and summed:

$$(x^2 - 4x + 4) + (y^2 - 6y + 9) = 25$$

or

$$x^2 + y^2 - 4x - 6y - 12 = 0.$$

EXAMPLE   Graph the circle having formula

$$x^2 + y^2 + 3x - 6y - 10 = 0.$$

To put this in the standard circle form, "complete the squares" of the x terms and of the y terms:

$$(x^2 + 3x + \quad) + (y^2 - 6y + \quad) = 10$$
$$[x^2 + 2 \cdot \tfrac{3}{2}x + (\tfrac{3}{2})^2] + (y^2 - 2 \cdot 3y + 3^2) = 10 + (\tfrac{3}{2})^2 + 3^2$$
$$(x + \tfrac{3}{2})^2 + (y - 3)^2 = 21\tfrac{1}{4}.$$

The center of the circle is at $C(-\tfrac{3}{2}, 3)$, and the radius is $\sqrt{21.25}$, between 4 and 5.

Exercise 7–49. Graph the circle

$$(x - \tfrac{5}{2})^2 + (y - 4)^2 = (\tfrac{7}{2})^2.$$

Exercise 7–50. Complete the squares and describe the circle

$$x^2 + y^2 - 2x + 10y + 10 = 0.$$

INTERPOLATION AND EXTRAPOLATION

The graph of $y = x^2$ in Figure 7–23 was drawn by plotting some known points: $(0, 0)$, $(\frac{1}{2}, \frac{1}{4})$, $(1, 1)$, $(1.5, 2.25)$, $(2, 4)$, $(2.5, 6.25)$, $(3, 9)$, $(4, 16)$, $(5, 25)$, $(6, 36)$, $(7, 49)$, $(8, 64)$, $(9, 81)$, and $(10, 100)$, and then sketching as smooth a curve as possible through them.

Once the curve is drawn you can use it to estimate values other than those that were originally plotted. If you use the curve between two plotted points, you are **interpolating**; if you use an extension of the curve past all the plotted points, you are **extrapolating**.

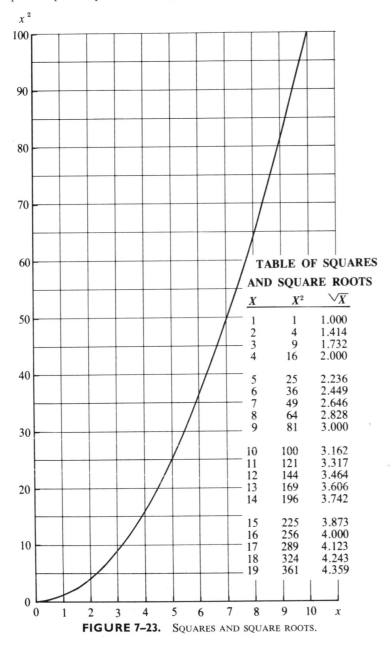

FIGURE 7–23. SQUARES AND SQUARE ROOTS.

TABLE OF SQUARES AND SQUARE ROOTS

X	X²	√X
1	1	1.000
2	4	1.414
3	9	1.732
4	16	2.000
5	25	2.236
6	36	2.449
7	49	2.646
8	64	2.828
9	81	3.000
10	100	3.162
11	121	3.317
12	144	3.464
13	169	3.606
14	196	3.742
15	225	3.873
16	256	4.000
17	289	4.123
18	324	4.243
19	361	4.359

EXAMPLE   Find $(3.5)^2$ from the graph in Figure 7–23. What is the y-coordinate for a point $(3.5, ?)$ if it lies on the curve? Vertically above $x = 3.5$, the curve appears to reach a height just above 12, perhaps 12.1 or 12.2. Check by finding 3.5^2:

$$
\begin{array}{r}
3.5 \\
\times 3.5 \\
\hline
1\,7\,5 \\
1\;0\,5 \\
\hline
1\;2.2\;5
\end{array}
$$

The graph was sketched from x values whose squares were known, but it can be used to find square roots, also, as we have already illustrated.

To estimate $(10.2)^2$ from the graph, you would need to extend the curve smoothly across the $x = 10.2$ grid line, then estimate the y-coordinate of the crossing. This process would be an extrapolation. Demographers often present their observations in the form of extrapolations: "If the birth rate continues to climb at the present rate . . . ," "At the present rate of urbanization . . . ," and the like.

You may find it much easier to spot trends and to interpolate in graphs than in lists of numbers in a table. You can enter the information given in a table on a cartesian graph to show at a glance where it is increasing, decreasing, positive, and so on. If the points you plot from the given table suggest a smooth curve, you can use the curve for interpolating.

Coordinate geometry and some interesting applications are covered in Chapter 8 of *Ideas in Mathematics* by Avron Douglis (W. B. Saunders Company, 1970).

REVIEW OF CHAPTER 7

1. Tell with examples exactly how to locate a point on a cartesian graph when you know its x- and y-coordinates.

2. If your pointer lands on a point of a cartesian graph, how can you find the coordinates of that point? Mention specifically what essentials of the cartesian grid you use at each stage.

3. What is the equation of a vertical line? of a horizontal line? of a line through the origin?

4. Write the equation of the family of lines through the point $(0, 4)$. What parameter determines the various lines of the family? Is the vertical line included?

5. Write the equation of the whole parameter family of lines having slope 5. Sketch 3 of these lines.

6. Find the distance between the points $(-1, -1)$ and $(1, 1)$. What is the formula for the distance between two points in the plane?

7. Describe the figure made up of all points whose coordinates satisfy the equation

$$(x - 2)^2 + (y + 18)^2 = 36.$$

8. Explain the relationship between the point of intersection of two lines and the simultaneous solution of the two linear equations.

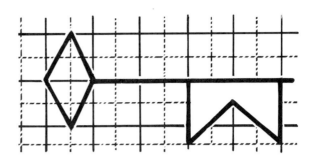

CHAPTER 7 AS A KEY TO MATHEMATICS

Perhaps you have seen a musician reading a musical score soundlessly but obviously responding as if he could hear what the notes represent. Or you may have seen someone reading blueprints for a building, picturing the completed structure as if it stood before him. Now that you have been introduced to analytic geometry you can understand how, with more experience, you might look at an algebraic equation and picture the figure made up of points with solution-coordinates. Already, when you see an equation of the form

$$Ax + By + C = 0$$

you think "linear"—"straight line." When you see an equation

$$Ax^2 + Ay^2 + Bx + Cy + D = 0, \qquad A \neq 0,$$

that is, an equation of second degree with the multipliers of x^2 and y^2 the same and the term in xy missing, perhaps you recognize it as describing a circle. Other second-degree equations represent ellipses, of which circles are a special case, hyperbolas, and parabolas. The graph of $y = x^2$ in Figure 7–23 shows part of a parabola.

The cartesian correspondence between number-pairs and points of a plane extends to a correspondence between algebraic equations and geometric figures. This correspondence makes the theorems of algebra available for use in geometry problems, and makes the ready intuition of geometry available to suggest solutions to algebra problems. In this way algebra and geometry are linked, each finding application in the other.

CHAPTER 8

The Modern Hiawatha
by George A. Strong*

He killed the noble Mudjokovis,
With the skin he made him mittens,
Made them with the fur side inside,
Made them with the skin side outside,
He, to get the warm side inside,
Put the inside skin side outside:
He, to get the cold side outside,
Put the warm side fur side inside:
That's why he put the fur side inside,
Why he put the skin side outside,
Why he turned them inside outside.

---

* From "The Song of Milkanwatha," in *The Home Book of Verse*, 5th Ed.
New York, Henry Holt, 1923.

TOPOLOGY

Figures That Stretch

You are curator of a collection of exhibits, including the 3-dimensional models shown in Figure 8–1. You have to label the various parts of each model in one of two ways:

either (1) label the faces and the vertices, as the tetrahedron (pyramid) is labeled in Figure 8–1,

or

(2) label the edges, as the cube is labeled in Figure 8–1.

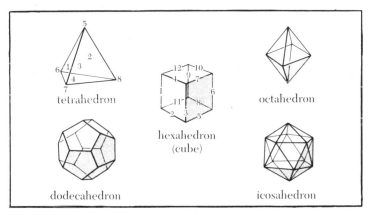

tetrahedron

hexahedron
(cube)

octahedron

dodecahedron

icosahedron

FIGURE 8–I. THE REGULAR POLYHEDRONS.

Which of the two methods will require fewer labels?

For the tetrahedron, method (1) requires 8 labels, since the tetra(4)-hedron(faces) has 4 faces and 4 vertices. Method (2) requires 6 labels, since inspection of the figure shows that it has 6 edges. Then, for the tetrahedron, method (2) requires 2 fewer labels than method (1).

Now make the same comparison for the hexahedron, and here use a solid model, since it is an easy shape to find. Incidentally, notice that for the purpose of counting faces, vertices, and edges, you do not need to insist on a perfect cube. A rectangular parallelopiped, such as the usual classroom or a cardboard box with the top in place, will do just as well. The hexa(6)-hedron(faces) has 6 faces, and it has 8 corners or vertices. As shown in Figure 8–1, the hexahedron has 12 edges. Then, for the hexahedron, method (2) requires 12 labels, 2 fewer than the $6 + 8 = 14$ of method (1).

Exercise 8–1. Make models of heavy construction paper for the tetrahedron, the hexahedron, and the octahedron (see Figure 8–2 for

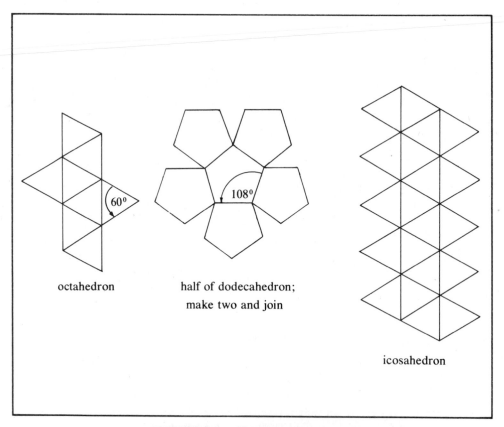

octahedron

half of dodecahedron;
make two and join

icosahedron

FIGURE 8–2. PATTERNS FOR PAPER POLYHEDRONS.

pattern), or find approximate models, sufficiently close to allow you to count faces, vertices, and edges.

Exercise 8–2. (Exercise 8–5 is an acceptable substitute.) Using the pattern shown in Figure 8–2, make a model of the regular 12-sided polyhedron, each face of which is a regular pentagon. The dodecahedron shape is sometimes used for paperweight desk calendars, one month for each face.

Exercise 8–3. (Exercise 8–6 is an acceptable substitute.) Following the pattern shown in Figure 8–2, make a model of a regular 20-sided polyhedron.

Lots of interesting and decorative semiregular polyhedrons, patterns, and a drinking-straw icosahedron are shown in Chapter 5 of *Mathematics, A Human Endeavor* by Harold R. Jacobs (W. H. Freeman and Co. Publishers, 1970).

Exercise 8–4. Referring to models or pictures, complete this table of numbers of faces F, vertices V, and edges E, and the difference $F + V - E$.

	F	V	E	$F + V - E$
tetrahedron	4	4	6	2
hexahedron	6	8	12	2
octahedron	8			
dodecahedron	12		30	
icosahedron	20	12		

As you have seen, a cardboard box of generally rectangular parallelepiped shape is close enough to a regular hexahedron for counting faces, vertices, and edges. Just what is a regular polyhedron, and which of its features are really necessary for an accurate count?

The **regular polyhedrons** are the solid figures bounded by congruent polygons having all their polyhedral angles (at their vertices) congruent. The five polyhedrons shown in Figure 8–1 are the only regular ones, as Cauchy proved in 1811.

In Exercise 8–4 you discovered for yourself **Euler's Theorem** for regular polyhedrons that

$$F + V - E = 2.$$

Euler proved this result, but it was known to Descartes a century earlier. Cauchy extended Euler's theorem to all simple polyhedrons, whether they are regular or not. **Simple polyhedrons** are polyhedrons without any tunnels, that is, without holes that go all the way through the solid.

Exercise 8–5. Use a general spherical mass such as a ball, a balloon, or a stuffed paper bag as basis for a deformed dedecahedron. Number the edges as you draw them, letter the faces as you enclose each one, and number each vertex with a roman numeral. Find $F + V - E$.

Exercise 8–6. Draw and number a deformed icosahedron on a generally spherically mass, as suggested in Exercise 8–5. Find $F + V - E$.

Exercise 8–7. Draw and number a deformed simple polyhedron on a generally spherical mass. Make some of its faces (deformed) triangles, some squares, some pentagons, and so on, so that the solid is definitely not regular. Show that Euler's theorem

$$F + V - E = 2$$

holds for the resulting polygon. (Save this model for Exercise 8–8. For this later use, it will be helpful if all but one face occupy a small, almost flat part of the surface, with any rough areas, neck of balloon, and so on, occupying the one face.)

Euler's theorem is unusual in your mathematical experience, and the proof is as unusual as the theorem. You may be familiar with proofs in plane geometry based on constructions. Here is a proof that starts with some destruction:

Let S be a simple polyhedron. Cut out one of its faces, leaving the bordering edges and vertices in place. Thinking of the rest as drawn on rubber, stretch it to form a flat "map," whose borders are the deformed edges of the removed face. Keep the same vertices and edges during the stretching process. Figure 8–3 illustrates the map for a dodecahedron.

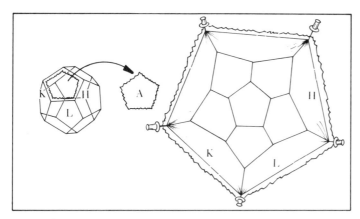

FIGURE 8–3. MAP PREPARED FROM DODECAHEDRON WITH FACE A REMOVED.

Exercise 8–8. Remove a face from your model of Exercise 8–7, and prepare a flat map from the rest of the polyhedron.

Although the polyhedron S is no longer assumed to be one of the few regular polyhedrons, we do imply some restrictions in the word "polyhedron": Each face must be a polygon. No more than one edge can connect two vertices. Exactly two faces meet at each edge. At least 3 edges meet at each vertex.

The requirement that S be a simple polyhedron is sufficient to assure that the flat map made by stretching S after removing one face is a connected network. A connected network is made up of polygonal faces such that every pair of vertices is connected through a chain of edges. (An example of a disconnected network is two triangles with no vertex in common.) The rest of this proof holds for any flat connected network.

Next, notice that if each individual face on the map is cut into triangles by joining one of its vertices to each of the others, there is no change in the quantity $F + V - E$, because the number of vertices stays unchanged, while each added edge adds a corresponding face. This step is illustrated in Figure 8–4 for the deformed pentagon H from Figure 8–3.

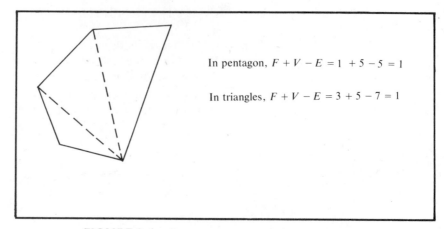

In pentagon, $F + V - E = 1 + 5 - 5 = 1$

In triangles, $F + V - E = 3 + 5 - 7 = 1$

FIGURE 8–4. PENTAGON FROM MAP CUT INTO TRIANGLES.

Exercise 8–9. Compute F, V, E, and $F + V - E$ before and after cutting into triangles the map faces shown in Figure 8–5.

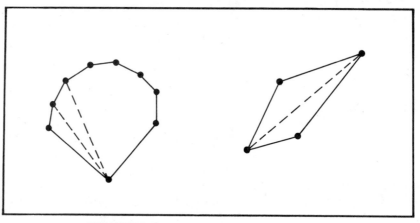

FIGURE 8–5. MAP FACES CUT INTO TRIANGLES.

Any triangle of the resulting map, all of whose faces are triangles at this stage, is called a "boundary" triangle if one, two, or all three of its edges are part of the boundary of the map, that is, an edge of the removed face. By experiment you can convince yourself that removing such a triangle does not change the value of $F + V - E$, provided the removal does not disconnect the network.

Exercise 8–10. Compute $F + V - E$ before and after removing boundary triangles from maps as indicated in Figure 8–6. Removing a

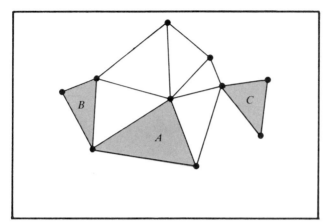

FIGURE 8–6. MAP SHOWING BOUNDARY TRIANGLES A, B, AND C.

triangle (A) having one boundary edge removes __1__ face, __0__ vertices, and __1__ edge. Removing a triangle (B) with two boundary edges removes __1__ face, ____ vertex, and ____ edges. Removing a triangle (C) with three boundary edges removes ____ face, ____ vertices, and ____ edges. In each case $F + V - E$ for the removed parts equals zero.

Although we do not prove it here, it seems plausible (and is the case) that if we remove border triangles in the right order, in particular saving for last those that can be removed by erasing one edge (type A in Figure 8–6), we can remove border triangles without disconnecting the network. Provided the network remains connected, the value of $F + V - E$ does not change.

Continue, then, to remove border triangles without disconnecting the network, and as this exposes new border triangles, remove those, until only one triangle remains. For that one triangle, $F = 1$, $V = 3$, and $E = 3$, so $F + V - E = 1$. Then $F + V - E = 1$ for the first map, since at each stage the sum has been preserved. But the map was formed by stretching the surface of the polyhedron after one face was removed. Adding in that face, we have for the polyhedron

$$F + V - E = 2. \quad \blacksquare$$

The sum is a constant, 2, for simple polyhedrons, and it is also a constant, but a different one, for polyhedrons of faces, edges, and vertices drawn on a torus, or tire shape, like that shown in Figure 8–7. In such polyhedral networks the sum $F + V - E$ is zero.

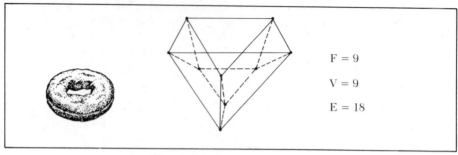

$$F = 9$$

$$V = 9$$

$$E = 18$$

FIGURE 8–7. A NETWORK ON A TORUS.

You can think of a torus as formed from a rough sphere by putting $p = 1$ tunnel through it. Figure 8–8 shows various "spheres" with $p = 0$, 1, 2, and 3 tunnels. It turns out that Euler's theorem can be generalized to show that a polyhedron on a sphere with p tunnels has $F + V - E = 2 - 2p$.

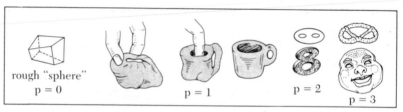

rough "sphere"
$p = 0$

$p = 1$

$p = 2$

$p = 3$

FIGURE 8–8. SPHERES WITH p TUNNELS.

You now have all the mathematical tools you need to prove that the five regular polyhedrons have the only face, vertex, edge patterns available for regular maps on a sphere.

Count the edges of a regular polyhedron in two ways. First, since the polyhedron is regular, each of its faces is a regular polygon, having a fixed number of sides, say p. Then the edges around each face number p, but pF, where F is the number of faces, counts every edge twice, once for each of the two faces that meet along that edge. Then

$$pF = 2E.$$

The second way of counting the edges is to count the number that meet at each of the V vertices. For a regular polyhedron the same number, say q, of edges come together at each vertex. If we form qV, we again count each edge twice, once for the vertex at each end of the edge. Then

$$qV = 2E.$$

From these counts, we can express F and V in terms of E, for substitution in Euler's formula.

$$F = \frac{2E}{p} ; \qquad V = \frac{2E}{q}$$

With these values for F and V, Euler's formula, $F + V - E = 2$, becomes

$$\frac{2E}{p} + \frac{2E}{q} - E = 2,$$

which can be divided by $2E$ and rearranged in the form

$$\frac{1}{p} + \frac{1}{q} = \frac{1}{2} + \frac{1}{E}.$$

Now suppose p to be 6 or larger. Then $1/q$ would be greater than $\frac{1}{2} - \frac{1}{6} = \frac{1}{3}$, which would make q less than 3. However, q represents the number of edges that meet at each vertex of a three-dimensional polyhedron, and hence must be at least 3. Similarly, from the fact that each face must have at least 3 edges, we can show that q cannot be 6 or larger. Then there are 9 available values for p and q: $\{3, 3\}$, $\{3, 4\}$, $\{3, 5\}$, $\{4, 3\}$, $\{4, 4\}$, $\{4, 5\}$, $\{5, 3\}$, $\{5, 4\}$, and $\{5, 5\}$. Only five of them make $\frac{1}{p} + \frac{1}{q}$ greater than $\frac{1}{2}$, and so provide face, vertex, edge choices for regular polyhedrons.

Exercise 8–11. Find which five choices of p and q make $\frac{1}{p} + \frac{1}{q}$ greater than $\frac{1}{2}$, and match each pair with its corresponding regular polyhedron, recalling the definitions of p and q. ∎

You will find an excellent presentation of these theorems together with related material on map coloring in *Excursions into Mathematics* by Anatole Beck, Michael N. Bleicher, and Donald W. Crowe (Worth Publishers, Inc., 1969).

In the process of reducing a polyhedral network to a manageable form, you have been led to alter the usual definition of a sphere. The usual definition requires that all points of the sphere be the same distance from the center, resulting in a completely symmetric solid. (President Eisenhower once complained that a certain man was a "spherical" pain in the neck, because he was a pain in the neck any way you looked at him.) The concept of a sphere has been generalized here to include stretching a rubber sphere without putting holes in it, nor bending it back on itself so as to enclose a tunnel. In altering the sphere according to this rule you changed many things, such as distance between points, enclosed volume, directions, angles, and areas, but you left invariant (unchanged) certain things you were interested in, notably $F + V - E$, and the number of tunnels $p = 0$.

The branch of mathematics called topology deals with this kind of change of geometric figures. It has been called "rubber sheet geometry," and the "reason you cannot put your pants on over your head." Figure 8–8 shows why topologists have been referred to as "mathematicians who cannot tell a doughnut from a cup of coffee," since both have $p = 1$, and so are equivalent (toruses) topologically.

You may have met a topological curiosity, the Möbius strip, perhaps in a magic show. If you join two ends of a paper strip without twisting the strip, you have a cylindrical shape with an outside surface and an inside surface, not different topologically from a length of pipe. Next, as shown in Figure 8–9, make a half twist in the strip before joining the ends.

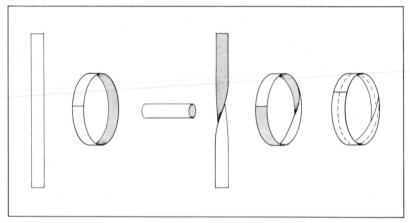

FIGURE 8–9. CYLINDER AND MÖBIUS STRIP.

Exercise 8–12. Form a half-twist Möbius strip and show by trying to paint just one side of the jointed strip that it has only one side. Show that it has only one boundary edge.

The Möbius strip you have made with a half twist is a one-sided surface. It can be covered or traversed completely without going over an edge.

Exercise 8–13. Make a paper figure like that shown in Figure 8–10. Show by trying to paint just one side that the figure has only one side.

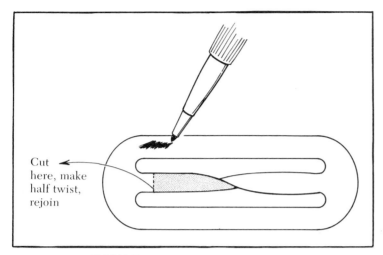

Cut here, make half twist, rejoin

FIGURE 8–10. A ONE-SIDED SURFACE.

Exercise 8–14. Some magic tricks are based on cutting Möbius strips. Cut a half-twist strip lengthwise as indicated by the dotted line in Figure 8–9. Demonstrate that the resulting surface has two sides. See what happens if you cut the resulting strip lengthwise.

Exercise 8–15. Form a full-twist Möbius strip and demonstrate by painting that it has two sides.

Exercise 8–16. Cut a full-twist Möbius strip lengthwise. Examine each link to see how many sides it has.

Exercise 8–17. Form a guess as to how many sides a one-and-one-half-twist Möbius strip has. Make one and confirm or reject your guess.

Topologists investigate, among other things, how many sides a figure has, and in particular whether it has an outside and an inside.

REVIEW OF CHAPTER 8

1. We usually think of mathematics as being "exact." Suppose someone looking over your shoulder at this chapter objects to stretching operations as inexact. Defend topology by stating clearly Euler's theorem for "spheres" and its generalization to spheres with p holes or tunnels.

2. Reread the proof of Euler's theorem for spheres and make an outline of it, that is, a series of titles for the separate steps in the proof.

3. Find three common objects that are topologically equivalent to a sphere. Find three objects equivalent to a sphere with one tunnel, and find two objects equivalent to a sphere with two tunnels.

4. Make a one-sided surface.

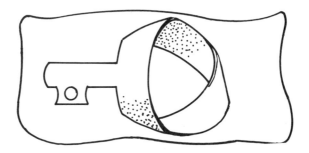

CHAPTER 8 AS A KEY TO MATHEMATICS

This short introduction to topology has been included partly to help you picture the wide scope of mathematics. As you have seen, even in a "geometry" with stretching, where measurement of distance between points loses its meaning, there are some dependable "invariants," such as Euler's $F + V - E$. Felix Klein (1849–1925), whose name was mentioned in Chapter 6 in connection with the Klein four-group, was a mathematician noted especially for his excellent teaching. He managed to reconcile conflicting specialties in mathematics by noting basic similarities. It was Klein who emphasized that different geometries had the common feature that each studied what was invariant under certain operations. In Euclidean plane geometry perpendicularity, parallelism, and so on are invariant under "superposition," moving a figure from one position in a plane to another. The constant $F + V - E$ is invariant under stretching.

CHAPTER 9

Where is the knowledge we have lost in
information

"The Rock" by T. S. Eliot

Statistics is not, properly speaking, a subset of mathematics, but
the two maintain a symbiotic relationship; statistics is depend-
ent on the mathematics of combinatorics and probability for
its foundation, and mathematics develops some of its most
important conjectures from the inductive conclusions of
statistics.

In everyday life we draw most of our conclusions inductively.
In fact, statements that are either True or False but not both
for every object in our experience must be considered rare.
However, we can put our statistical conclusions on a mathematical and, hence, logical basis, so that a probabilistic statement
("There is a 40% chance of rain.") has a clear and consistent
meaning, as invariable as any mathematical definition.

STATISTICS

Decisions on Partial Evidence

You are the principal of a junior high school. The Citizens for Innovation in Education want a new independent study program (call it *N*, for "new"), but the Board of Education wants to continue the standard (call it *S*) classroom method, unless *N* can be shown to be definitely superior, because *N* would require hiring two more teachers. To determine whether one method is better than the other, and if so which one, you plan to try method *N* for one year, as an experiment. How should you plan the experiment so that it will give you reliable enough results for a reasonable *decision*?

How can you measure the effectiveness of an educational method? One *index* (or measure) of effectiveness you might use would be the increase in score between an achievement test administered at the beginning of the trial year and another administered at the end.

Suppose the score at the beginning of the year was 7.0 and after a year with the standard method *S* the score was 8.0. Then you could say that the difference $8.0 - 7.0 = 1.0$ represented the effectiveness of method *S*. See Figure 9–1. If the *before* and *after* scores for method *N* were 7.2 and 8.5, the index for *N* would be $8.5 - 7.2 = 1.3$. Since the *N* index would be higher than the *S* index, you would say that *N* was more effective than *S*. **EXAMPLE**

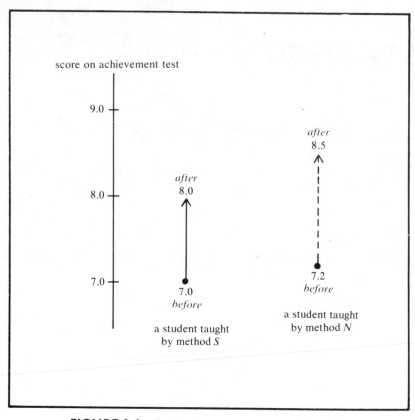

FIGURE 9–1. INDEX OF EFFECTIVENESS OF S AND N.

If a student taught by method S had a *before* score of 7.0 and an *after* score of 8.0, then the effectiveness of S for him would be $8.0 - 7.0 = 1.0$. If a student taught by method N had a *before* score of 7.2 and an *after* score of 8.5, the index of effectiveness of method N for him would be his improvement, $8.5 - 7.2 = 1.3$.

FEATURES OF A STATISTICAL EXPERIMENT

Your experiment to compare S and N as educational techniques has features typical of statistical experiments: a *decision* to be made, some *index* of the effects to be compared, a *population* concerned in the decision, a *sample* giving incomplete information about the population, and a *distribution* of index values over the population, some values more probable, some less probable. We shall highlight each of these features in turn.

1. The purpose of the experiment is to help you come to a **decision**, and a wrong decision either way carries penalties.

EXAMPLE   If you decide to adopt method N, but method S is really just as good, the penalty is the cost of hiring two extra teachers.

Exercise 9–1. Suppose N is definitely superior but your experiment is poorly designed and does not bring this out, causing you to decide in favor of S. What "penalties" might ensue?

Exercise 9–2. You set up an experiment to decide whether to change your laundry soap to a new low-phosphate product. Discuss the penalties of a wrong decision for the standard soap S, and the penalties of a wrong decision for the new soap N.

Exercise 9–3. Discuss the penalties of a wrong decision either way in the trial of an alleged lawbreaker.

2. You have in the difference between the *before* and *after* scores a quantitative (numerical) **index** of the effectiveness. This index is a bit artificial, and it does not tell everything about how S and N compare— with respect to student, teacher, or parent approval, for instance—but it does measure something closely related to effectiveness of a technique. It is not always necessary to have a numerical index. Often a subjective judgment is all that is necessary. For instance, it might be adequate to have a student subject in the experiment check which method had seemed more effective in his own case.

EXAMPLE

To a limited extent cost can be used as an index of quality, number of votes as an index of popularity, grades as an index of ability. Each of these indices has deficiencies that may make it unsuitable for a particular experiment. The closer the index is linked to the final decision, the better results the experiment can have.

Exercise 9–4. Suggest an alternative index in the education experiment and show where it is better and worse than the rise-in-score index.

Exercise 9–5. Discuss 3 possible indices for the soap experiment of Exercise 9–2.

Exercise 9–6. Choose one of these indices, tell what it is supposed to measure, and mention any deficiencies it has. Why is it used?

> GNP, gross national product
> Dow-Jones stock average
> I.Q., intelligence quotient
> A, B, C, D, F, or Pass/Fail college grades

3. One of the most important features of the experiment is imaginary! In comparing the two education methods S and N, you imagine a whole **population** of changes of score, some for girls, some for boys, some for good readers, some for poor readers, some for students actually involved in the experiment as subjects, some not yet in school. To get good, reliable

results from your experiment you need to imagine accurately! That is, you need a clear conception of just what population you are making a decision for.

EXAMPLE   Perhaps you have decided in advance that only students with a B average will be considered for the independent study program N. If so, this forms an important restriction on the population and should be included in a written description of that population. Other questions you should answer explicitly at this time are: Will method N be used only for certain courses, such as English, history, social studies, or will it be used for everything from shop to band? Will students apply to take method N, so that only students aggressive enough to make such an application belong in the population? Will there be any special considerations that would rule certain students out of the population, such as a record of discipline problems, or the recommendation of previous teachers that extra supervision seems necessary?

Exercise 9–7. A statistician takes a market survey for the manufacturer of a new kitchen appliance. Write three sentences that describe the population to be sampled. (For example, need the survey cover children?)

Exercise 9–8. You plan the output to be exported from a garment factory in Hong Kong. What do you need to know about the population of future customers?

4. The **sample** is just as real as the population is imaginary. A crucial part of the experiment is relating the one to the other. The picture of the population must be built up until it is quite clear and complete, so that the sample can be drawn from the right population. So far as possible, the sample should be drawn **randomly** from the whole population; this means that each member of the population should have an equal chance to be chosen to be a sample subject. Randomization is used to avoid **bias** in the sample, that is, an emphasis on part of the population at the expense of other parts. A familiar way of selecting a random sample is to draw lots.

EXAMPLE   The education experiment comparing methods S and N is to take place during a single school year, so the sample cannot be drawn randomly with respect to time: A student who hasn't yet entered school does not have any chance to be chosen for the sample during the trial year. In such a case, the only way to minimize bias is to foresee special effects due to the year chosen for the experiment and neutralize them in choosing the sample. For instance, there may be some students who are especially aggressive in demanding a change from method S to method N, or one group of incoming students may have transferred from a rigidly formal school. Ideally, such special categories of students should have a chance to be included as subjects in the sample in proportion to their estimated place in the whole population. The sample subjects from each special category should be chosen randomly from the whole category. For example, to draw 12 sample subjects from the "aggressive" students, place number chips or enrollment cards for all the aggressive students in a box. Stir, and select one chip or card blindly, stir again, and continue until 12 have been selected.

Exercise 9–9. Show how the pictured population affects the proper choice of sample by contrasting the problems of getting a representative sample in these two cases: First, suppose you are trying to reach a decision between a standard soap S and a new low-phosphate product N for your own laundry at home. Second, suppose you are offering home-extension advice for laundries throughout the county.

Exercise 9–10. Does the mail on a congressman's desk give him a random sample of opinions in his constituency?

Watch for bias in samples offered to you as persuasion.

People are often persuaded by the sample of "all my friends" or "all the people I've heard about" or "the neighbors," forgetting that these samples are probably biased as to socioeconomic level or hobby-interest or student status, and these biases may be important to the effect under study. Another common bias is introduced if we limit study to subjects who volunteer or who happen to be nearby for other purposes. **EXAMPLE**

Exercise 9–11. A miner takes a *sample* of ore to be analyzed. What is the *population*? Look up the terms "high-grading" and "salting" as used in mining.

Exercise 9–12. Discuss a social science example of population-and-sample of your own choosing. For instance, you might discuss taxation in relation to owning real estate or other property, or the Electoral College system, or ways of choosing regents for higher education in the various states.

Exercise 9–13. You sell apples at a roadside stand. It is your practice to conceal bad apples in the bottom of the barrels, covering them with a layer of good apples. Use the technical terms "population," "sample," "bias," and "random" to describe this process.

5. The concept of a whole spread, or **distribution,** of values for the index is characteristic of statistics problems. You do not imagine a "correct" index for teaching method N that tells exactly how effective it is

for every student. The question "What is the effectiveness of an educational technique?" does not have any one clear answer in the sense that the arithmetic question "Which whole numbers are between $5\frac{1}{2}$ and $10\frac{1}{2}$?" does. Corresponding to the whole population of students you imagine, you would expect lots of different responses to an educational method like N, some responses large (much improvement), some small.

Among all the responses there may be central ones that occur relatively frequently, and outlying ones that occur relatively infrequently. The solid curve in Figure 9–2 depicts a possible distribution of responses to method S. Most of the area under the curve is concentrated near the response 1.0, indicating that the most frequent responses are thought to be near 1.0, an improvement of 1 point. A small proportion of the population of students would make almost no improvement (response near 0), and an even smaller proportion would make negative improvement (slip back). A very small proportion of the population might improve more than 2 points. In this way, although we have no single answer to the question "How effective is method S?" we may be able from experiment, past experience, and so on, to get an idea of the distribution of responses to S, how spread out they are, and where most are concentrated.

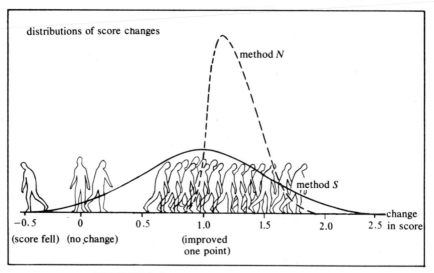

FIGURE 9–2. DISTRIBUTIONS OF SCORE CHANGES.

The solid curve shows responses to method S. For the whole population of students a large proportion of responses are fairly near 1.0. Over half the area under the curve lies within one-half point of 1.0, to show that over half the population shows a response to method S of between 0.5 and 1.5 points. The proportion of students improving by more than 2.5 points is very small, probably negligible. This is shown by the fact that the area under the distribution curve tapers down to a negligible amount to the right of 2.5. A very small proportion of the population may be expected to show negative improvement by getting a lower score after method S than before. This is shown by the small area under the distribution curve to the left of 0.

The dotted curve shows a possible distribution for N responses, with relatively frequent responses around 1.3 points.

The dotted curve in Figure 9–2 shows a possible distribution curve for N-responses. It suggests that responses to method N are concentrated around an improvement of 1.3 points, rather than just 1 point, and that the responses are not quite so spread out as the S-responses, the proportion of outlying responses being somewhat smaller.

It would be useful to farmers to know when to expect the first killing **EXAMPLE** frost of winter. This is not a question that has a mathematical answer, though. The best the Weather Bureau can do is supply a distribution of first-frost dates from past records, showing which ranges of dates are relatively most probable, which very improbable, and so on.

Exercise 9–14. Discuss what a teacher might mean by the "distribution" of grades on the last test.

Exercise 9–15. Discuss what a political scientist might mean by the "distribution" of wealth in a country.

Exercise 9–16. Discuss what a reformer might mean by promising a "redistribution" of land.

Exercise 9–17. Suppose the National Safety Council predicts 700 traffic deaths over a holiday weekend. Does that mean they are pinpointing 700 particular individuals who are likely to die in traffic accidents? In what way do previous records supply a picture of the *distribution* of traffic fatalities over weekends, commuters' hours, holidays, wet days, and so forth?

Before learning more about experiments and how to use them in coming to a decision, you may want to review the vocabulary of the last few pages:

> *decision*—The purpose of the experiment is to gather evidence to use in reaching a decision. Ordinarily the evidence is not conclusive as in a mathematical proof but instead indicates which is *probably* the best decision.
>
> *population*—The objects about which the decision is to be made constitute the population. Often it is at least partly imaginary or unavailable for experiment.
>
> *sample*—The actual experiment does not involve the entire population to be decided about, so the population is represented by a sample drawn at random from the population to avoid bias. A random system such as drawing lots is used so that each object in the population has an equal chance to be drawn for the sample.
>
> *index*—Some measurement is taken for each sample object in the experiment.

distribution—We assume that there is some index measurement (whether it can actually be observed or not) for each object in the population, and that the index measurements are distributed among various values, some values occurring relatively frequently, other values occurring relatively infrequently.

Decision: whether to use background music in a store

Index: dollar volume of business in a day

Population: volume of business on all days the store will ever be open

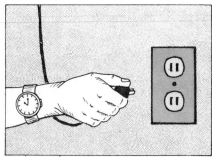

Sample: hours of time on comparable days, some with music, some without

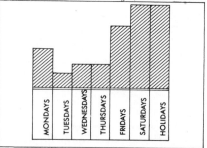

Distribution: how the dollar volume of business varies from day to day

FIGURE 9–3. TYPICAL FEATURES OF A STATISTICAL EXPERIMENT.

FIGURE 9–4. STATISTICAL TERMS IN CARD GAMES.

Each time playing cards are shuffled and dealt, the cards have a *distribution* among the players' hands. Each player's cards represent a *random sample* from the deck, which represents the *population* of cards. A *decision* about what to bid is called for in some games. The decision might rest on an *index* reflecting the strength of the hand, say, the number of cards in the strongest suit, perhaps weighted to let high cards have more effect than low ones.

PLANNING THE EXPERIMENT

Next, we are going to introduce a particular statistical test, called the sign test, and show what kind of experiments it is appropriate for.

We have chosen the sign test partly because it does not require any calculations at all! Toward the end of most statistical experiments comes a stage called **reducing the data.** This amounts to studying the records of the experiment after it has been performed to spot trends. Often those records are largely numerical, and the search for trends proceeds arithmetically, with the computing of various averages, differences, and so on. It was once common to use the word "statistics" to refer to the comparatively unimportant process of reducing the data and summarizing the results in tables and graphs. Perhaps because modern computers have simplified the reduction of data, we now see the statistician's most important contribution to be at the planning stage, before the experiment is run! It is at this stage that the statistician can advise the experimenter on

all the experimental designs appropriate to his material, what the requirements for each one are, and how the data should be analyzed for each one.
The statistician can be especially helpful in recommending a sample size
that will minimize the cost of the experiment in time and money while
meeting the experimenter's limitations on the probable cost of a wrong
decision (more on this later).

The Sign Test

One test a consulting statistician might suggest for your comparison
of S and N is the sign test. To make the sign test appropriate for analyzing
the results, the experiment must be designed so that subjects are taken in
matched pairs. In the education experiment each sample student should
be matched with another who is as much like him as possible with respect
to their potential response to methods S and N. This does not mean they
must look alike, of course, since presumably their appearance would not
bias them toward S or toward N, but they should have similar past achievement records, similar attitudes toward participating in the experiment
(both apathetic, for instance, or both especially interested). If you think
boys may be especially prejudiced toward N, say, then boys should be
paired with boys and girls with girls.

By a random process, such as flipping a coin, assign method S to one
member of each pair, method N to the other. At the end of the experimental year, record a + for a sample pair if the student taught by method
N improved more than the student taught by method S. Record a − if
the student taught by method S improved more. Record a 0 in case of a
tie, that is, if the difference between the *before* and *after* scores was the
same for both students in the pair.

EXAMPLE   Suppose David and Jonathan were paired in the experiment. They
flipped a coin and David got "heads," which by prearrangement assigned
him to the new method N, and Jonathan to method S. David had *before*
and *after* achievement test scores of 7.1 and 8.2. Jonathan had scores of
7.2 and 8.5. Since Jonathan improved more than David (8.5 − 7.2 >
8.2 − 7.1), record a −.

Mark and Sam were assigned to methods S and N, respectively, by
flipping a coin. Mark had *before* and *after* scores of 6.5 and 8.0; Sam had
scores of 7.5 and 9.0. Since each improved by 1.5 points, record a 0 for a
tie.

To apply the sign test, count the total number of signs, plus or minus.
This is just the number of pairs included in the experiment minus the
number of tie results, so that you count just those pairs that showed more
improvement by one method than by the other. Then count whichever is
smaller, the number of + signs or the number of − signs. If almost all

of the signs are $+$ or almost all of the signs are $-$, then the smaller count will be quite small and *probably* the two methods are different in their effects. If there are about the same number of $+$ signs as $-$ signs *it seems doubtful* that there is any real difference in effectiveness between S and N.

We say "probably" the two methods are different, or "it seems doubtful" that there is any real difference. The table in Figure 9–5 shows in the

Level of Significance α

Total	0.05	0.10	0.25	Total	0.05	0.10	0.25	Total	0.05	0.10	0.25
1	—	—	—	36	11	12	14	71	26	28	30
2	—	—	—	37	12	13	14	72	27	28	30
3	—	—	0	38	12	13	14	73	27	28	31
4	—	—	0	39	12	13	15	74	28	29	31
5	—	0	0	40	13	14	15	75	28	29	32
6	0	0	1	41	13	14	16	76	28	30	32
7	0	0	1	42	14	15	16	77	29	30	32
8	0	1	1	43	14	15	17	78	29	31	33
9	1	1	2	44	15	16	17	79	30	31	33
10	1	1	2	45	15	16	18	80	30	32	34
11	1	2	3	46	15	16	18	81	31	32	34
12	2	2	3	47	16	17	19	82	31	33	35
13	2	3	3	48	16	17	19	83	32	33	35
14	2	3	4	49	17	18	19	84	32	33	36
15	3	3	4	50	17	18	20	85	32	34	36
16	3	4	5	51	18	19	20	86	33	34	37
17	4	4	5	52	18	19	21	87	33	35	37
18	4	5	6	53	18	20	21	88	34	35	38
19	4	5	6	54	19	20	22	89	34	36	38
20	5	5	6	55	19	20	22	90	35	36	39
21	5	6	7	56	20	21	23				
22	5	6	7	57	20	21	23				
23	6	7	8	58	21	22	24				
24	6	7	8	59	21	22	24				
25	7	7	9	60	21	23	25				
26	7	8	9	61	22	23	25				
27	7	8	10	62	22	24	25				
28	8	9	10	63	23	24	26				
29	8	9	10	64	23	24	26				
30	9	10	11	65	24	25	27				
31	9	10	11	66	24	25	27				
32	9	10	12	67	25	26	28				
33	10	11	12	68	25	26	28				
34	10	11	13	69	25	27	29				
35	11	12	13	70	26	27	29				

FIGURE 9–5. CRITICAL VALUES FOR THE SIGN TEST.

(Sign table courtesy of W. J. Dixon and F. J. Massey, Jr.: *Introduction to Statistical Analysis*, 3rd Ed. New York, McGraw-Hill Book Company, 1969.)

"level of significance α" how probable a count of + and − signs is if S and N are really equally effective. For example, opposite a total of 11 and under the significance level 0.25 the table lists a critical value of 3. This is a highly condensed way to say that the following "splits" of + and − signs are so unbalanced that they appear only about 25 per cent of the time if the methods are equally effective:

> ③ and 8 (3 +'s and 8 −'s *or* 3 −'s and 8 +'s)
> 2 and 9
> 1 and 10
> 0 and 11

The level 0.25 means that the frequency of such unbalanced splits by chance alone if S and N are equal would be not more than 25 out of 100.

A few exercises on the mechanics of applying the sign test will help you follow a further explanation of how Figure 9–5 is constructed and interpreted.

EXAMPLE Suppose you began with 50 pairs of students, two pairs of which record zero difference in improvement. Then the total count of signs is $50 - 2 = 48$. Of those, suppose 35 are +'s, so that 13 are −'s, reflecting more improvement for the N member of the pair in 35 cases out of 48. To see whether 13 is fewer − signs than you would expect from chance alone if S and N are equally effective, consult Figure 9–5. Opposite the total count of 48, under the level of significance $α = 0.25$, find the critical value 19. Yes, a 13 to 35 split is more unbalanced than the listed 19 to 29 ($= 48 - 19$) split, so conclude that S and N are significantly different in effect at the 0.25 level. The 0.25 level means that in the long run you would conclude that a significant difference existed when actually S and N were alike only 25 percent of the time.

EXAMPLE Suppose you record 11 + signs and 20 − signs. Then the total is 31, with a critical value from Figure 9–5 of 11 at the significance level 0.25. Since the smaller number of signs, 11, is less than or equal to the critical value 11, conclude that there is a significant difference between S and N. In the long run such a conclusion would be drawn by chance alone about 25 times out of 100 even though S and N were equally effective.

Exercise 9–18. Suppose you record 25 + signs and 20 − signs. What do you conclude from the sign test, according to the critical value for the 0.25 level of significance in Figure 9–5?

Exercise 9–19. The Home Team and the Visitors play 10 baseball games, each carried to a non-tie conclusion. How many games does one team have to win to convince you at the 0.25 level of significance that one team is significantly better rather than just lucky?

Exercise 9–20. A doctor administers cold vaccine to one of each pair of patients, matching them according to their estimated susceptibility and exposure to colds. He tests 83 pairs of patients. In 48 cases the patient receiving the vaccine reports definitely fewer colds than the untreated patient paired with him. How does the doctor write his conclusion based on the sign test?

Exercise 9–21. The last 18 times I dropped my bread, it landed jam-side down 8 times, jam-side up 10 times. Should I conclude from the sign test that bread is more likely to land one way than the other?

Exercise 9–22. Twenty volunteers compare two soft drinks. Sixteen of them prefer Brand *X*. Is Brand *X* significantly different from the competitor?

The Level of Significance and the Power of the Test

How is statistics related to mathematics? Making decisions on incomplete evidence seems far removed from the deductive logic (Chapter 3) required in a mathematical setting.

Statistical conclusions are of the sort: "This quantity of smoke would be improbable if there were no fire," or "That quantity of smoke would be no more than we would expect even if there were no fire." It is in finding what is probable that mathematics enters. The **probability** of a certain response is the relative frequency of that response in the population. The probabilities of various responses can often be calculated exactly, assuming certain facts about the population.

For example, the critical values for a total number of signs equal to 6 in Figure 9–5 can be calculated as follows: First, assume that S and N are equally effective in the population, that is, that +'s and −'s are equally likely to occur. What possible arrangements are there for samples of six signs? Figure 9–6 shows the 64 possible samples, arranged according to how many + signs they have. Of the 64 possible different samples, 20 of them have the split of signs exactly balanced, 3 and 3; 30 of them have 2 of one sign and 4 of the other; 12 of them have 1 of one sign and 5 of the other; 2 of them are all one sign. Then the relative frequency in the population of samples of 6 signs having the unbalanced split 0 and 6 is 2 out of 64, or $\frac{2}{64}$, or about 0.03. This relative frequency in the population is called the *probability* of a sample of 6 signs having the smaller number of signs equal to 0, still assuming +'s and −'s are equally likely in the population.

The probability of a split as unbalanced as 1 and 5, that is, either 1 and 5 or 0 and 6, is $12 + 2$ out of 64, or $\frac{14}{64}$, or about 0.22. The prob-

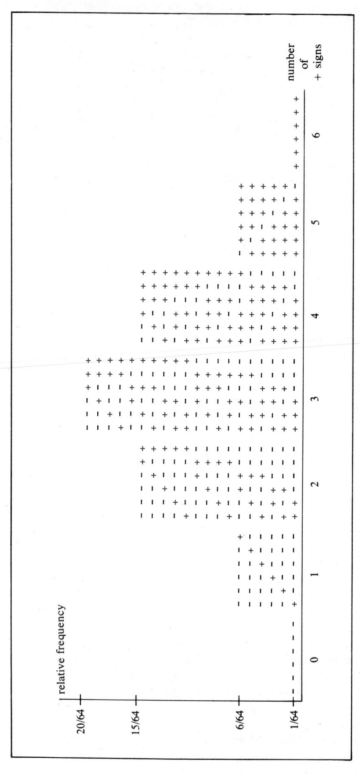

FIGURE 9–6. THE 64 POSSIBLE SAMPLES OF 6 + SIGNS AND − SIGNS.

There are 64 possible arrangements of + signs and − signs in samples of 6 signs. If + signs and − signs are equally likely in the population, then each of these 64 possible samples is equally likely in drawing random samples of 6 signs. From this, the relative frequency of samples having a particular number of + signs can be calculated as the proportion of 64 that they constitute. Then if +'s and −'s are equally likely, that relative frequency is the probability of drawing such a sample.

ability of a split as unbalanced as 2 and 4 is 30 + 12 + 2 out of 64, or about 0.69. That is why Figure 9–5 lists as the critical value at the significance levels 0.05 and 0.10 a smaller sign of 0, for a 0 and 6 split occurs with relative frequency only about 3 times out of 100 in a population in which +'s and −'s are equally likely, while a split as unbalanced as 1 and 5 occurs with probability 0.22, which is more than 0.05 and 0.10. The critical value at the significance level 0.25 is given as 1, because such an unbalanced split occurs with probability 0.22, which is less than 0.25, while a split as unbalanced as 2 and 4 occurs with probability 0.69.

Since the level of significance α is the probability of an error, the error of claiming significance when actually there is none, why not use $\alpha = 0.05$ or even smaller in every case, to minimize this probability of error? The reason is that we must compromise with the probability β of making a second kind of error, which is to decide there is no real effect when actually there is one. For example, suppose in the education experiment that one method is so much more effective than the other that if a + or a − could be recorded for every pair in the population there would be three times as many + signs as − signs. Then to show the relative frequency of various numbers of + signs in samples of 6 signs, each + sign in Figure 9–6 must be replaced by 3 + signs. This replacement is shown for a few cases in Figure 9–7, and then the relative frequencies are shown to a smaller scale.

How **powerful** is the sign test against the **alternative** having 3 times as many + signs as − signs; that is, how probable is it that this alternative population will produce a sample of 6 as unbalanced as the critical value in Figure 9–5? The answer depends on the level of significance. At a significance level of 0.05 or 0.10 the critical value in Figure 9–5 is 0. How frequent are samples this unbalanced in the alternative population illustrated in Figure 9–7? In Figure 9–7 one sample has 0 + signs and 729 samples have 6 + signs, so the relative frequency of a critically unbalanced split of signs is 1 + 729 out of 4096, or about 0.18. This means that the probability of finding a significant difference in this case is only about 18 out of 100, or equivalently, 82 times out of 100 a sign test on samples of 6 signs from such a population would not have a significantly unbalanced split. If the sign test is conducted at a significance level of 0.25, then the critical split is 1 and 5, according to Figure 9–5. Samples as unbalanced as this occur in the alternative population with relative frequency 1 + 729 + 18 + 1458 out of 4096, or about 0.54. This means that on the average about 0.54, or slightly over half of our samples of 6 signs, would show this much unbalance and indicate a decision that the two methods are not equally effective. The probability β of wrongly deciding for equal effectiveness would be $1 − 0.54 = 0.46$.

To summarize, then, since the statistical decision is made for the whole population from just a sample, there are risks that the sample will be atypical, leading to a wrong decision. The **significance level** α is the probability that although in the population there is no significant effect, a (correctly drawn) random sample will by chance exhibit such unbalance that the decision will claim significance. The **β risk** is the probability that

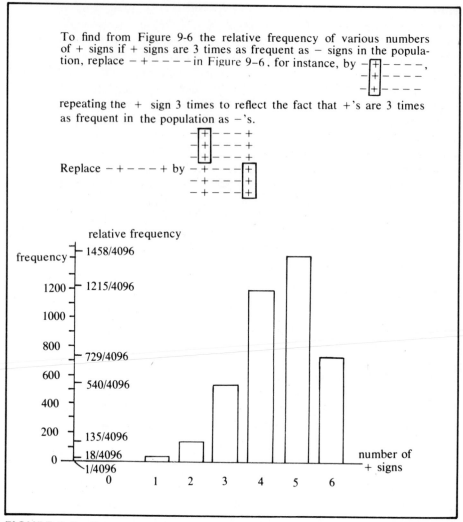

FIGURE 9–7. RELATIVE FREQUENCY OF SIGN SPLITS IN SAMPLES OF 6 SIGNS WHEN + SIGNS ARE 3 TIMES AS FREQUENT AS − SIGNS.

although there is some significant effect in the population, a random sample will by chance be well enough balanced so that a decision of no significance is reached.

α = probability of claiming significance (because of an atypical sample) when there is none in the population

β = probability of deciding no significance (because of an atypical sample) when an alternative is true. The "**power of the test** against that alternative" is $1 - \beta$.

In setting the level of significance of any statistical test, we have to compromise between the probabilities α and β of the two kinds of error.

In considering the relative importance of the two kinds of error, we consider the penalties of each kind of wrong decision. To decrease one of the probabilities of error without increasing the other, we have to increase the sample size. The probabilities of the two kinds of error for the sign test described in Figures 9–6 and 9–7 are too large to be very satisfactory: $\alpha = 25\%$ and $\beta = 46\%$. The reason both probabilities are so large is that the sample is small, deliberately chosen at only 6 to make the counting of frequencies easy.

FIGURE 9–8. The level of significance, the power of a test, and the cost of experimenting (continues Figure 9–3).

Suppose a couple of dedicated statisticians decide to have six children   **EXAMPLE** in order to test whether they are equally likely to have boys or girls. They stand a 25% chance of producing a family with no girls or no boys or just one girl or just one boy, even if girls and boys are equally likely for them. On the other hand, suppose their genetic balance is such that girls are three times as likely for them as boys. They would still stand a 46% chance of a fairly balanced split (at the 25% level of significance) of boys and girls, two of one sex and four of the other, or three of each.

Suppose the Board of Education is determined not to adopt method N unless it is very clearly superior, because of the additional cost. If the citizens' committee is not so persuasive, perhaps because they do not have a direct effect on your job, you will lean in favor of making N prove itself. In such a case you might choose α to be 0.10 or even 0.05, since you are very anxious to minimize the probability of concluding significant difference if there is none. The level of 0.25 reflects more interest in picking up a difference if any exists, avoiding the error of concluding that they are about the same in effect.

Exercise 9–23. Suppose the firemen are outside your door with axes and spray foam in readiness. What are the penalties for a wrong decision on: "Where there is so much smoke there must be some fire"?

Sample Size

A statistical consultant can help you choose a sample size for your experiment at a given level of significance to make your test have the power it needs against various alternatives. To lower the probabilities of the two types of error, you increase sample size. This can introduce another cost consideration. You may have to balance the cost of a wrong decision either way with the cost of increasing the size of the experiment.

Exercise 9–24. Proponents predict that the new method N will produce on the average 0.5 more improvement than method S. This can be tested by the sign test by adding 0.5 to each S improvement and comparing with the paired N improvement. What does the level of significance 0.25 mean in this case?

Exercise 9–25. Flip a coin 16 times, recording $+$ if it is heads, $-$ if it is tails. Pool the results of many such experiments. According to Figure 9–5, about 1 in 4 experiments should have a 5 to 11 split in signs or worse, in the sense of less balanced. About 1 in 10 experiments should have a 4 to 12 split or worse. About 1 in 20 experiments ($\alpha = 0.05$) should have a 3 to 13 split or worse.

Exercise 9–26. Flip two coins, recording $+$ if either or both are heads, $-$ only if both are tails. An "experiment" consists of 16 tosses of the two coins. Use the sign test to determine whether the chances of a $+$ or a $-$ are about the same. Actually, a $+$ is three times as probable as a $-$, since there are three ways of getting a $+$ for each way of getting a $-$; namely, two heads, heads on the first coin with tails on the second, and heads on the second with tails on the first. By pooling the results of many such experiments, get an empirical (by test only) estimate of the power of the sign test against this alternative when the level of significance is fixed at 0.25.

Exercise 9–27. Dr. Sickby is trying to devise a test to pick out people who are especially prone to heart attacks as compared with the general population. He takes a few readings from each patient. How stringent should he make his test: Should he decide a subject is a potential heart-attack risk if a single reading seems abnormal? In that case which risk would be high, the probability α of deciding the patient is a heart-attack risk when he is not, or the probability β of deciding he is normal when he is really a heart-attack risk? Suppose he decides a subject is a potential heart-attack risk only if almost all the readings are abnormal. Then which probability would be high, α or β? If he wants to minimize both probabilities he must take more readings, which may be expensive or even physically hard on some patients. Balancing the probabilities of error and the cost of sampling—x-rays, biopsies, and so forth—is a very real problem in medical testing.

Selecting the Sample

As you see, the mechanics of applying the sign test are very simple. You do not need multiplication, square roots, or even addition, but only counting. Other tests may require more computing, but still be better than the sign test for some experiments. It is important to study the assumptions on which each test is based to see which ones are appropriate for a problem. The sign test is based on the assumption that the two members of each pair are alike in all factors that might affect their response to one method as compared with the other. What factors should be allowed for in this pairing?

1. You may need to pair high achievers together and low achievers together on the basis of past scores or teacher recommendations. You must judge this from your experience. Do high achievers have more to gain from a switch to independent study methods than do low achievers? Do low achievers have more to gain because they have been insufficiently challenged by the standard method?

2. Does native intelligence, measured approximately by an I.Q. score, affect the student's potential response to one method over the other?

3. Should you take the student's attitude into account, pairing students who have been demanding a change or who resent authority?

4. You may want to pair students according to a subjective impression of a teacher, that is, the teacher's prediction of how the student will respond to the two methods.

5. You may want to pair students according to their record of library use, or according to whether they have reference books at home, or according to their main field of interest. You may want to pair them according to some index of parental approval of the project or according to how eager they are to participate.

As you consider each relevant factor, let it help you develop a clearer picture of the population to be included in your decision. For example, as

Fridays and Saturdays seem to have about the same number of shoppers and a similar weekend atmosphere. We can pair them to use in the sign test.

Do we have to compare whole days, or can we compare similar hours?

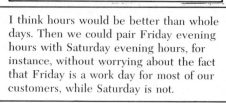

I think hours would be better than whole days. Then we could pair Friday evening hours with Saturday evening hours, for instance, without worrying about the fact that Friday is a work day for most of our customers, while Saturday is not.

Suppose we have worked out a sample of paired hours so that the two hours in a pair are similar. Then we flip a coin to see which hour gets the music. Right?

Right, and we record a+ if we did more business during the hour with music, a− if we did more during the hour without.

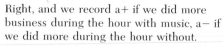

Wait a minute. The sound system is going to cost us something. I don't think we should put it in unless it makes us take in quite a bit more per hour.

We can allow for that. Just record a+ if the volume of business is a dollar more with music and a− if it is not as much as a dollar more than the hour without music. Any more questions?

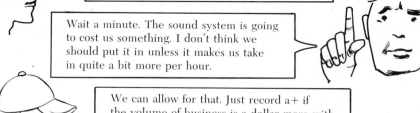

Yes. You know nobody's going to want music Monday morning. Everybody is crabby Monday mornings.

You might be right, but then again, maybe they need music all the more. I'd include Mondays in our sample to be sure you're right. Then if people don't seem to like music on Mondays but do buy more to music on other days, it may still be worth it to put in the sound system and simply unplug it Monday mornings. Then the "population" would include only Tuesday through Saturday.

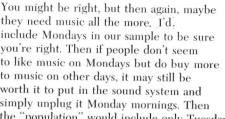

FIGURE 9–9. PLANNING AN EXPERIMENT TO DECIDE WHETHER TO INSTALL BACKGROUND MUSIC IN A STORE (CONTINUES FIGURES 9–3 AND 9–8).

you consider point 1, about level of past achievement, you may come to realize that only the high achievers will be included in the eventual population of students affected; that is, that low achievers will not in any case be permitted method N, even if it is adopted for the school. If so, such a conclusion must be reflected in your sample to make it representative. The performance of low achievers under the two methods should not influence your decision about whether to adopt one or the other for high achievers.

Exercise 9–28. Suppose you decide that I.Q. score is essentially irrelevant with respect to comparing S and N. Then do you need to be sure the members of each pair have similar I.Q. scores?

Exercise 9–29. Suppose that in considering point 3, you feel from your experience that the attitude of the student has an appreciable effect on which method works better for him. Show how to reflect this in your sample, or show how to redefine the population, in case you decide that some attitudes prohibit improvement under one method. If you make such an adjustment in the population, show how to adjust the sample accordingly.

Exercise 9–30. Point 4 suggests a different statistical problem, that of determining whether a given teacher or guidance counselor can predict better than by chance alone which students will improve more by method N, say, or pass the next course in a sequence, or earn more than $10,000 a year within 10 years after graduation. Describe a sign test for this problem.

You may want to use one student as *both* parts of a pair, by teaching him history, say, by method S, but English by method N. Then at the end of the experiment you can compare his improvement in history with his improvement in English, as indicated by subject *before* and *after* scores. This kind of "pairing" is very desirable if it holds constant factors that might affect response. Exceptions can arise and part of your role as experimenter is to anticipate them. For example, suppose a student's attitude or achievement differs markedly between history and English. Then if you feel attitude and achievement level are important to the response to S and N, you cannot make a pair of his marks in these two subjects. Perhaps you feel that a student can handle one or two subjects on an individual basis N, but that more than two independent study programs will compete too strenuously for his time. In that case, redefine the population according to whether you will restrict program N to one class alone, always to history, to English, and so on. Then make the sample representative. If the final program would involve all subjects, but you think the number of subjects included may be relevant, then you cannot pair a single student's performances on two subjects, but must pair two students with a similar curriculum.

Exercise 9-31. Does the index have to be numerical, or could the sign test be applied to a subjective rating for each pair, say a + if the pair of students considered N the better method and a − if they considered S better?

PROBABILITY

Probability, especially precipitation probability, has become a familiar feature of weather forecasts. Suppose the forecast for your area says the probability of rain today is 80%. What does the figure 80% stand for? The forecaster takes into account such information as time of year, temperature, recent barometer readings, weather in neighboring areas, winds, and so on. Then he estimates that of all days sharing similar indications, about 80 out of 100 such days have a measurable amount of rain.

You have seen that an experimenter tries to assess the penalty of a wrong decision either way. An example arises when a precipitation probability is used in making a business decision.

EXAMPLE   You are a new car dealer, with 15 new cars on display outdoors. You estimate that it costs you about $1 to have each car driven into shelter for the night, counting labor, possibility of dented fenders, and so on. On the other hand, if a car is left outdoors and it gets rained on, you estimate that it costs you about $10 to touch up its polish. Suppose the forecast says there is a 50% chance of rain tonight. Should you have the cars driven into shelter?

First, suppose you have the 15 cars sheltered on 20 occasions when the forecast gives the rain probability as 50%. Then the cost is a dollar for each of the 15 cars each of the 20 times, or

$$\text{cost of sheltering} = \$1 \times 15 \times 20 = \$300.$$

Second, suppose you do not have the cars sheltered those 20 times and 50% of those times it does rain. Then the cost is $10 for each of the 15 cars each of the 10 rainy nights, or

$$\text{average cost of not sheltering} = \$10 \times 15 \times (50\% \text{ of } 20) = \$1500.$$

The comparison of the two cost risks indicates that the better decision is to have the cars sheltered whenever the rain probability is given as 50%. Now suppose the rain probability is 20%. For 20 such nights we have the comparison

$$\text{cost of sheltering} = \$1 \times 15 \times 20 = \$300$$

$$\text{average cost of not sheltering} = \$10 \times 15 \times (20\% \text{ of } 20) = \$600.$$

Again, the decision to shelter the cars whenever the forecast is 20% rain probability minimizes your financial risk.

Exercise 9–32. Suppose the rain probability is given as 10% for 20 nights. Show that the cost of sheltering the cars equals the average cost of not sheltering them, so that the two risks balance exactly.

Exercise 9–33. If you carry your umbrella to the city with you, you pay 25¢ to check it in a locker. If it rains and you do not have your umbrella with you, you pay $1.00 to have your suit pressed. Compare the probable costs of 20 trips with or without umbrella when the rain probability is 30%. Which decision minimizes cost risk?

Exercise 9–34. In Exercise 9–33, compare the probable costs of 20 trips with or without umbrella when the rain probability is lowered to 20%. Show that the minimum cost decision changes between 30% and 20% rain probabilities.

Independence; Correlation

On Shrove Tuesday each year, ladies in Olney, England, compete in a pancake race with ladies in Liberal, Kansas, in the United States. Presumably, Olney and Liberal are parts of such different weather patterns that the forecaster for Olney would not even take into account the weather in Liberal when estimating the rain probability in Olney, and, similarly, the weather in Olney has no apparent relation to the weather in Liberal. Suppose that the rain probabilities for next Shrove Tuesday are 60% for Olney, and 50% for Liberal. What is the probability that both towns will have rain? In a long run of, say, 100 such predictions, Olney expects 60 rainy days and Liberal expects 50 rainy days. Since the weather patterns are independent, we assume that the 60 rainy days in Olney will be distributed about evenly between Liberal's 50 rainy days and Liberal's 50 dry days, that is, about 30 of them falling on rainy days and 30 of them on dry days. This means that about 30 days of 100 will be rainy in both towns. The multiplication of the probabilities is the way we express in arithmetic what we mean by **independence.**

$$
\begin{array}{ccc}
\text{rain} & \text{rain} & \text{probability of} \\
\text{probability} \times \text{probability} = & \text{rain in both} \\
\text{in Olney} & \text{in Liberal} & \text{towns} \\
60\% & 50\% & 30\%
\end{array}
$$

In contrast to events that are independent are events having a high **correlation.** Suppose that two small towns are located near each other in the same valley, ordinarily sharing wind and temperature conditions. Notice that while the weather in one town does not *cause* the weather in the other, both depend on the same movements of air, and so forth, so that rainy days in one town tend to occur with rainy days in the other. We say that rain in the one town has a high positive correlation with rain in the other. Suppose two other towns have a natural barrier between them, such as a high mountain, tending to produce dry days in one if winds from the direction of the other are relieved of their water vapor in rising over the mountain. The rain incidences of the two towns in such a case are negatively (but highly) correlated.

Weather probabilities make a good example from which to learn the terms, but the ideas of independence and correlation are certainly not restricted to meteorology. They are important, with wide application.

Exercise 9–35. You choose a boy and a girl to lead a game at a party. If you choose the boy at random from four boys, Matt, Marco, Lucius, and Jack, and the girl at random from three girls, Esther, Ruth, and Martha, write the 12 possible combinations. The probability of picking any one boy, say Marco, is one out of four, or 25%. What is the probability of picking any one girl, say Martha? What is the probability of picking any one of the combinations, such as Matt-Esther? If you picked the boy at random but then let him choose the girl he liked best, the choices would no longer be _____.

Exercise 9–36. Suppose that 50% of a certain high school graduating class go on to college. Suppose that 90% of graduates from the high school who go to college go to a state university. About what percentage of that graduating class go to a state university?

Exercise 9–37. My car has a fuel system that is about 95% reliable (works 95 times out of a hundred) and an ignition system that is about 80% reliable. How reliable is the car, taking it to be mainly the combination of the two?

Exercise 9–38. (Also read Exercise 9–37.) I have sold my patented secret weapon to the Sinister Syndicate, but they complain that it works less than half the time. Why is this, since my weapon contains 8 essential independent parts, each 90% reliable?

The Normal Distribution

A weather forecaster wants to find what the usual low temperature is for January 16. Checking the records for his forecast area, he finds an average low of about +18.8°F. How useful is this average for predicting the low on next January 16? This depends on how variable the temperature is. Does it always fall within a few degrees of the average in the forecast area, or are some Januaries much colder or much milder? To answer these questions, the forecaster studies the whole distribution of lows for January 16.

If the forecaster has complete records for many years, he will probably find a distribution of lows that can be approximated closely by the normal distribution illustrated in Figure 9–10. The lows will probably prove to be distributed fairly symmetrically around their average of 18.8°, with about

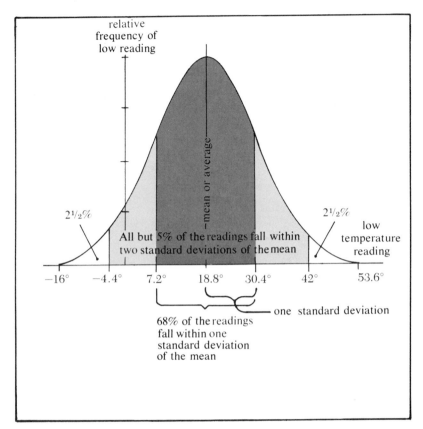

FIGURE 9–10. A NORMAL DISTRIBUTION OF TEMPERATURE LOWS.

the proportions shown falling close to 18.8°, and far from 18.8°, respectively. Theoretically, lows as high as 60° or as low as −20° are possible, but improbable. The center or average value is called the **mean,** and the unit measure of variation about the mean is **one standard deviation.**

The importance of the normal distribution as an approximation to nature stems from a fact expressed mathematically in the Central Limit Theorem. This theorem shows that under very general circumstances a quantity is approximately normally distributed if it is the sum of many independent random effects. We can think of the temperature low as the sum of a great many random effects that are independent from each other, sometimes reinforcing, sometimes cancelling each other, but not consistently correlated with each other.

A psychology major records the number of seconds it takes each of **EXAMPLE** 200 rats to run a maze. Presumably, the time each rat takes represents the sum of many independent effects. Probably the normal distribution closely approximates the experimental one.

Figure 9–11 shows recorded lows for each day of January, 1966, in Manhattan, Kansas. The curved solid line connects the mean lows, averaged over the previous 30 years. The dotted lines connect points one, and two standard deviations away from the mean, again computed from the variability shown in the previous 30 years.

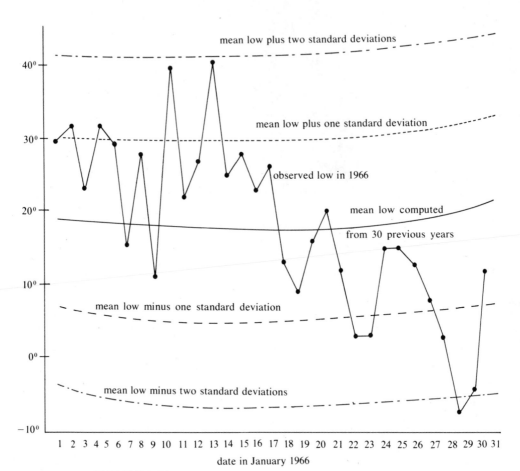

FIGURE 9–11. AVERAGE LOW TEMPERATURE. (Data courtesy KSU Weather Station, Manhattan, Kansas.)

Exercise 9–39. Count the 1966 daily lows in Figure 9–11 that fell within one standard deviation of the mean. Compare with the proportion of the 31 readings expected to fall within one standard deviation of the mean according to Figure 9–10. What percentage of the 31 readings fell outside two standard deviations?

Exercise 9–40. The standard deviation on some test scores was 5.1 points. The mean was 75 points. The instructor wants to assign a letter grade of "C" to grades within one standard deviation of the mean, letter grades "B" and "D" to those within two standard deviations but not within one, and grades "A" and "F" for the others. Show how many points constitute an "A," a "B," and so on.

You may have encountered an instructor who "graded on the curve." This probably meant that he followed a scheme like that in Exercise 9–40. It is hard to justify the use of the normal distribution for grades on a test, because of the difficulty of defining the population and the sample meaningfully. Are the grades to be considered a random sample from the population of grades that all students having that instructor would make if given the same test? Or, is the population all students taking the particular course at that school, or in the state or country, or over the past 10 years? Can the grades of one class represent the population reliably by constituting a random sample of sufficient size? Is the distribution of grades in the population well approximated by a normal curve? One common shape for distributions of grades is "bimodal," that is, it has no single center of concentration, but two separate points around which the grades cluster. A bimodal distribution of grades might arise on an addition-and-subtraction quiz if some of the students did not follow the instructions well and added in every case instead of subtracting on about half the problems. Then their grades would cluster around 50%, depending on how accurately they worked the addition problems. The other students would have grades that clustered near 100%, depending on how accurate their addition and subtraction were.

As usual in statistics, choices depend on what use is to be made of the experimental results. If a bimodal feature in a distribution of grades can be used to diagnose students' deficiencies in arithmetic skills, then it is undesirable to combine the two modes into an artificial average having no natural interpretation.

Accuracy and Precision

In Figure 9–12, picture (a) shows a hose nozzle adjusted so as to give a wide spray centered on a birdbath. In this case the aim is accurate, but

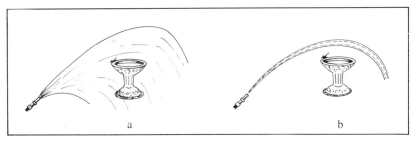

FIGURE 9–12. ACCURACY AND PRECISION.

not precise. In picture (b) the nozzle is adjusted to give a narrow stream, which, however, misses the birdbath. In this case the aim is precise but inaccurate. In a distribution, the mean gives a measure of central location, while the standard deviation measures dispersion (spread-outness). Both are important to prediction. To know how much importance to attach to an average, you need to know whether it represents most readings pretty well (whether it is precise), or whether it is merely the center for widely scattered readings (accurate but not precise).

Exercise 9–41. You are testing laundry detergents again. You notice a difference in color and perfume among different samples of the same product. This leads you to question whether each product is variable in washing effect within brand. If it is, what effect could this have on your comparison?

Exercise 9–42. Criticize comparisons like "Fourth generation Americans are more law-abiding than immigrants" and "Blondes have more fun than brunettes" on grounds that the difference in means, if any, is small compared with the standard deviation.

PARAMETRIC TESTS

Figure 9–13 (a) shows a family of normal distributions, all having a common standard deviation but having different values for the parameter: the mean. The family of normal distributions in Figure 9–13 (b) have the fixed mean zero, but they differ in the parameter: the standard deviation.

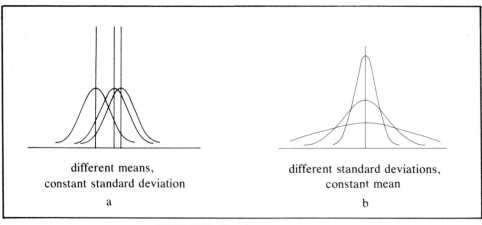

different means,
constant standard deviation

a

different standard deviations,
constant mean

b

FIGURE 9–13. FAMILIES OF NORMAL CURVES.

(For the parameters slope m and y-intercept b for a straight line, see Chapter 7.) Many statistical tests are based on the assumption that the index under study has a normal distribution with one or both parameters unknown. The question to be decided may be: "Are the sample values about what could be expected by chance if the unknown mean is $18°$?" or "Are the sample values compatible probabilistically with a hypothetical standard deviation of 0.7 points?"

Returning to the education experiment, suppose we believe that the responses of students to the standard classroom method S follow a normal distribution, mean 1.0, standard deviation, say, 0.5. This means that we expect about 68% of all students in the whole population to improve by about 1.0 ± 0.5 points on *before* and *after* achievement tests when they are taught by method S. If we believe that method N also produces a normal curve of responses with standard deviation 0.5, then we can make a parametric test to see whether the means are probably the same, or whether one method is superior (has a higher mean). If the assumptions of normality hold, a smaller sample size will give the same probabilities of both kinds of error as the sign test. The sign test is one of several tests called "nonparametric," because they can be used when the experimenter cannot specify the distribution even as to shape (with various parameters yet to determine).

For more about the concepts and philosophy of chance and statistics, read pages 151–178 of *Mathematics in the Modern World*, Readings from Scientific American (W. H. Freeman and Co. Publishers, San Francisco, 1968).

REVIEW OF CHAPTER 9

1. Discuss these terms as they arise in statistics:

> decision
>
> index
>
> distribution
>
> population
>
> sample

2. Suppose that you are conducting a statistical experiment to determine whether subjects learn mathematics best at one particular time of day or whether there is no significant "circadian" effect. What are the two kinds of errors and some of the penalties associated with each?

3. In question 2, which subjects should be in the sample? Exactly how might they be selected?

4. Discuss randomization in sample selection, using question 2 for illustration.

5. What does a forecast of "30% probability of measurable precipitation" mean?

6. A neighbor likes her new toaster very much, and I am considering buying one of the same brand. I would like to know how good the producer's quality control is, that is, how small the standard deviation of toaster performance is. How would that help me decide how much weight to give to my neighbor's approval of her toaster?

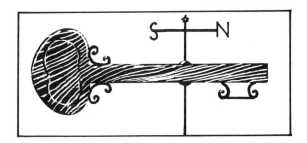

CHAPTER 9 AS A KEY TO MATHEMATICS

Usually mathematicians draw their conclusions (theorems) from "evidence" (a hypothesis) that is not only sufficient but overwhelming. In fact, a mathematical proof shows that the hypothesis is sufficient to imply the conclusion.

In your daily life you are faced with decisions (conclusions) that must be drawn from incomplete evidence. If you must make important decisions, you try to balance the alternatives. In this you function like a statistician. Unable to examine the whole population of objects involved, the statistician theorizes about the population, and checks his theories by analyzing random samples. If he finds the samples very unusual for the supposed population, he modifies his theories about the population. Mathematicians can contribute at this stage, through the theory of probability, by finding exactly how unusual certain classes of samples would be, given the supposed population. Inexact as other stages of the process may be, the probability part is exact. Consider a statement like: "If paired samples are randomly drawn from a population of pairs with no real difference, then the probability of 6 or fewer minus signs in the sign test is 0.25." Such a statement is not guesswork, but can be calculated mathematically.

CHAPTER 10

"Well, now, you hear me, Epaminondas, you be careful how you step on those pies!"

. . . .

And then—and then—Epaminondas *was* careful how he stepped on those pies!
He stepped—right—in—the—middle—of—every—one.

Epaminondas and His Auntie
*by Sara Cone Bryant**

Much of what we have learned in other chapters comes together here: the algorithm approach to solving problems, abstraction and mathematical models, good notation and tabulation, format in reasoning and in calculation. Flowcharting may deepen your understanding of algorithms and help you analyze problems, even if you do not use a computer.

---

* Houghton Mifflin Company, 1938, pages 15–16.

COMPUTERS

Why and How to Talk to Them

WHY COMMUNICATE WITH COMPUTERS?

What Have Computers Done for You Lately?

If you attend a large university, computers probably registered you for this course. Computers can be fed such information as

what courses you want
what courses you have prerequisites for
what courses you need to graduate
what courses are offered this term
what scheduling problems you have because of part-time jobs.

Then the computer can be fed a system of priorities, for instance:

Reject anyone not correctly registered as a student.
Close courses scheduled for ordinary classrooms at 30 students.
Close lecture-hall classes at 300 students.
If demand exceeds space in sections planned, alert the department office to see whether they can staff another section.
Decide conflicts in favor of those who are required to take the course, and against those for whom it is an elective.

Exercise 10–1. Review the college admissions procedure in Chapter 1, page 10. Adapt it to the problem of registering students in classes: Have each student rank the courses he would consider taking this term according to how eager he is to take them. Suppose there is a quota for each section of each course. How might conflicts be resolved? (One possibility is listed among the possible priorities fed to the computer.)

Exercise 10–2. Review the transportation problem covered in Chapter 1, page 12. How could a computer assist the University Committee on Inner Space in allocating a fixed amount of small-classroom time and a fixed amount of lecture-hall time among the Mathematics Department, the Physics Department, and the Statistics Department? Develop

(1) an index to be *maximized* based on how many students are able to get the courses they want, or

(2) an index to be *minimized* based on how many students are prevented by a conflict from getting the courses they want.

Notice that allocation of space is actually more complicated than the situation covered in Exercise 10–2. Lecture-hall time, for instance, does not come in a lump sum, but is divided into class hours on various days of the week. A student who cannot meet a 10 o'clock class Monday, Wednesday, and Friday may be able to meet an 8 o'clock class on Tuesdays and Fridays.

Exercise 10–3. Find out whether computers are involved in the registration process at your college. If so, visit the computing facility or arrange a lecture by personnel who can explain some of the features of the program.

Exercise 10–4. Find out what computers help figure your college bills and visit the facility, or arrange to have someone from the Business Office give a talk about business computers. Does the Business Office share computer time with other users?

Exercise 10–5. Write some subjective responses to registration procedures you remember. To what extent were they computerized? What constructive suggestions can you offer? If registration is to be computerized next term, what directives would you suggest for guiding computer decisions; that is, what should be the list of priorities for resolving conflicts?

Exercise 10–6. Few computer registration programs allow students to choose their instructors. How could computers advise on faculty appointments, granting of tenure, raises in pay or rank, and so on?

Exercise 10–7. Comment on the academic advisory system at your college. Does each advisor know the catalog requirements, such as required hours for graduation, required courses for your major, prerequisites for each course in your program? How would you react to advice from a computer that had the contents of the college catalog stored in its memory? Would it make a difference to you whether the computer answered your questions in a human voice or in print? Would you prefer advice from a computer-assisted human counselor?

One thing computers have done for you lately is assign you the kind of nicknames that come naturally to them: the kind they find familiar and easy to remember, that is, numbers. You have an area code for telephone communication, a zip code for mail communication, a Social Security number for identification, probably a car license plate number, a library card number, a bank account number, a student identity card number, and some credit card account numbers.

Some of your numerical names indicate where computers have entered your life. In communications, the telephone network relies heavily on computers. Zip codes represent an attempt to make postal service eligible for computer assistance. Satellites that carry much of our communication these days, including TV news transmissions, depend on computers in their design, placement, and use.

Exercise 10–8. Find out what information is contained in your zip code. Which of the five digits contain which information?

Exercise 10–9. Find out from the front of a telephone directory how to get the telephone number of someone in a distant city. What provision is made for billing in case of wrong numbers on long distance direct-dialed calls?

In transportation, computers book reservations around the world via satellite, determine landing and takeoff schedules at airports, optimize bus schedules, help designers of city transit systems, and even help the police protect you from auto theft.

Exercise 10–10. Find the nearest transportation or reservation agency that uses computers: airlines, bus lines, travel agencies, or motel-hotel chains. If possible, learn how well the employees or customers think their computer program works. Do they have serious complaints?

Exercise 10–11. Find out how serious the auto theft problem has been in your state or county in recent years. To what extent are authorities hampered by slow information processing or poor information flow across jurisdictional boundaries?

Exercise 10–12. Describe peak traffic situations you know about, perhaps stadium traffic for a football game. What is done to alleviate congestion? What might be done? How could computers help?

Computers are especially prominent in your financial life, figuring your bank balance, deciding your credit rating, putting you on mailing lists for various advertisers, totaling your credit card bills.

In the larger economic picture, computers behind the scenes help in production decisions: what products should be offered to consumers, and at what prices?

Exercise 10–13. Try to read from a check the number of the checking account. Read the information printed by a computer on a bill—for gasoline, for instance.

Exercise 10–14. Relate any errors that have cropped up in your financial affairs handled by computers. Did you have any trouble drawing human attention to the problem? Do you have any financial dealings that would be improved by computer assistance?

Closely related to computer studies of consumer acceptance of commercial products are computer projections leading to political and policy decisions. Computers can project from early samples a picture of how several possible nominees might stand on election day. Of course, these predictions are no better than the samples they are based on, which may be too small or too early to be representative.

Exercise 10–15. Recall TV coverage of an election. Was reportage limited to votes already counted, or was much of the coverage devoted to projections?

A major use of computers in planning is simulation. To design a new type of rapid-transit vehicle, a new highway complex, an economic program, or a farm support program, the planners may be able to simulate a proposed design on a computer. Then they can give the computer data on city transportation needs, on highway use, on economic variables, or on agriculture statistics, and let the trial-and-error phase of design be played out on the computer, instead of in real life.

An important and growing use of computers is in medicine. A cardiac patient in a hospital intensive care unit may be linked to computers that monitor his vital functions continuously and alert personnel if there is any significant change. The computer may even initiate corrective measures directly. A researcher studying a contagious disease can have a computer summarize statistics from far-flung reporting stations and so chart the course of the disease and predict its future course. He may even be able to correlate the incidence of the disease with other factors, thereby providing a clue to the mechanism of contagion. A researcher seeking a plant substitute for sugar, say, may have tests on thousands of plants summarized and stored in a computer memory in such a way that he can quickly retrieve information according to any new, promising criterion— results on all the 3-lobed leaves found on Dessert Island, for instance.

Figures 10–1 through 10–5 illustrate some uses of computers.

T. T. Command No. 427: One of Bill Amstein, Jr's. prize Hereford bulls.

The record work the Amsteins are doing with the American Hereford Association is for gain and records on their cattle. It's all done by computer. The records come back to the Amsteins with production of dams and sires.

"The print outs," he stresses, give "us some information we can really use, whereas we just don't have time anymore to do it by hand."

Courtesy of *Kansas Country Living*, March, 1972, page 11.

Until recently, the vibration strain on the linkages imposed a limit on the speed of mowing machines. Then a model of the process was simulated on a computer, resulting in redesign of the linkages.* Now the limit on speed is the ability of the tractor operator to stay in the seat, and he must wear a seat belt!

FIGURE 10-1. COMPUTERS IN AGRICULTURE.

* By Professor Hugh Walker of Kansas State University, for Hesston Corporation, Hesston, Kansas.

Minicomputer Makes Oil Pipeline Safer, More Economical

Almost hidden in the massive and extensive array of equipment in what may be the world's largest and longest heated crude oil pipeline in the world, a tiny minicomputer is making significant contributions to both safety and economy of the entire operation. The Varian 620/i minicomputer is part of a solid state supervisory control system, at the point of origin of a 174-mile 20-inch pipeline moving heavy crude oil

Help for Jet-Age Joe . . . Automatic computer-aided passenger check-in devices for air travelers are being developed by Sperry Rand Corporation's Univac Division to ease crowded conditions at terminals as air ~~travel triples~~ in the 1970's. The s~~ervices~~ include credit

"Talking" Computers Highlight New Banking Advances at Automation Conference

An RCA "talking" computer at Cherry Hill, N. J., is telling bankers here in a poised, almost human voice details on simulated checking and savings accounts. At the same time, other executives press a typewriter-like keyboard and watch banking data flash on a TV-like screen from an RCA Spectra 70/46 time sharing computer in ~~Los~~ Angeles.

The RCA ~~,~~

Computer Helps Company Map Utility Poles, Pipelines . . . A T~~elephone~~ ompany is mapping power lines in far-away Wyoming and tracing pipelines in the Alaska wilderness with a computer that never leaves Antonio. Tobin Aerial Survey Co. uses an IBM 1130 system and a plot device to generate these specialized maps. The information is gathe

FIGURE 10–2. COMPUTERS IN BUSINESS. (Courtesy *Software Age*, March through June, 1970.)

THE COMPUTER AS SORTER

Returning football scouts once sorted and re-sorted by hand their notes on athletes to find how they ranked in yardage gained, in passing, in kicking, and in blocking. Now, scouts and recruiters use the computer for sorting. Recruiters for professional teams use computers to keep track of statistics on promising college athletes.

POLE VAULTING

FIGURE 10–3. COMPUTERS IN SPORTS.

Professor Hugh Walker of Kansas State University has worked out a mathematical model of pole vaulting, so that coaches can simulate the entire vault on a computer and optimize the many variables, such as length of the fiberglass pole, flexibility of the pole, types of handholds, and takeoff factors.

Hospital Computer Service

A radiologist prepares an analysis of x-rays on a television-like terminal located in a hospital laboratory. The device is connected to McDonnell Automation Company's computers in Peoria, Ill. The hospital data processing system provides administrative and patient care information.

Computer Helping Improve Care for Mentally Ill

A computer is helping trai professionals provide faster and ter treatment for the mentall and retarded of Texas. Psychia now receive rapid patient inf tion from the computer that them diagnose illnesses an' scribe and maintain pr treatment. Ps~' greater inform fore about the needs of ment state's far-flung schools.

Soon, the IBM each patient fron discharge, record: his treatmen progr

Computerized Medical Data System Announced

A new computer system for transmitting medical and health care claims information electronically to and from doctors' offices was recently unveiled. Installation of the system will start in May.

The system, developed jointly by Western Data Products and Blue Shield of Virginia, will cut health care claims costs (for Blue Shield and Medicaid) by over $1 million in Virginia alone and could $60 million a

TRACK USE OF OVER 1,000 DRUGS . . . A computerized system for stocking more than 1,000 drugs and intravenous solutions is helping pharmacists and doctors improve tient care at Shadyside Pittsburgh. Shad... vate i-- 360 t track ability up to pharmac ing is re lengthy ing names and chlori Under cards are intrave-

DIAGNOSTIC TESTS . . . Opening of a unique Biomedical Screening Cen ter, equipped to perform basic medical diagnostic tests rapidly and accurately was recently announced by Martin Kaplan, president of Biometric Systems Inc., Jericho, N. Y., the parent com pany.

With the cooperation of life insur ance companies, the Center expects to concentrate upon physical examination rance applicants. All functions supervision of qualified testing car

Computer Helps Scientist Study Inner Ear

A mathematician at IBM's Los Angeles Scientific Center has programmed a computer to simulate the intricate workings of a portion of the human ear. Dr. Alfred Inselberg, who has spent 11 years on the has created a mathematical the inner ear that could ialists learn more about orks and might suggest r certain types of hear-

l, already valuable in es about inner ear one of eight such re selected for demon IBM pavilion at To-

The simulation ds of thousands of

FIGURE 10–4. THE COMPUTER IN MEDICINE. (Courtesy *Software Age,* March through June, 1970.)

FIGURE 10–5. MISCELLANEOUS COMPUTER APPLICATIONS. (Courtesy *Software Age,* March through June, 1970.)

INSTANT INVOICING

A computer system has recently been designed (by Professor Walker, K.S.U.) for filling the daily reorders for garment patterns. Clerks tear off an identification tab each time they sell a pattern, and the tabs are sent in as reorders. The identification numbers are fed to a computer, which activates a mechanical selector, meanwhile automatically preparing the invoice and bill, and updating the inventory record at the pattern center. The selector withdraws the ordered patterns from among some 5000 bins of patterns in different styles and sizes.

LIE DETECTOR RESEARCH BY COMPUTER . . . Lie detector tests may be made more accurate because of computer-based research at Delta College, University Center, Mich. Prof. William Yankee, academic dean of the college and a nationally known polygraph expert, is using an IBM computer to more accurately analyze factors he feels are critical in determining deception by a polygraph test subject.

Speeding the Word

Yesterday's criminal could lose himself in the ? ment message transmission. But not any longer. Us CLETS—the California Law Enforcement Telecommuni than 450 urban and rural law enforcement agencies in seconds to retrieve criminal records stored in c

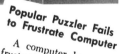

Popular Puzzler Fails to Frustrate Computer

A computer has overcome the frustrations of "Instant Insanity," a popular puzzle that requires the ...s in a ...e there ...

Musical Work Composed With Aid of a Computer

An 11-minute modernistic mus composition written by a Greek-b composer with the aid of a comp was performed by a symphony chestra for the first time in the February 7, "ST/48 for 48 Play a 1962 work of Iannis Zenakis played by the Pro Arte Symp Orchestra, under the directic Eleazar de Carvalho at the H University Playhouse, Hemp L.I., N.Y.

Composer Xenakis, a mat tician, engineer and architec holds a degree in civil engir from Ecole Polytechnique, / also has studied computer pr ming. His musical compositi produced in two steps. F wrote a computer program b a set of mathematical laws (ing random variables. This p in turn, directed a random generator program stored w computer which picked

COMPUTER METER-READER STUDIED . . . A computer could read your electric, water, or gas meter automatically through your telephone line without ringing the phone or preceptibly tying up the line. To test such a system, trials involving up to 150 homes at Holmdel, N. J., will be started in a month or two by Bell Telephone Laboratories. More extensive trials will be held in other locations ... 1970

Au some telepl comp ment to be cient Labs of th

Electric Communications to Aid Traffic Safety Flow

Motorists of the future may have a number of electronic communications systems to improve safety and traffic flow, a General Motors research executive told delegates to the 17th International Scientific Congress on Electronics recently. Edward F. Weller, head of the Electronics and Instrumentation Department of GM Research Laboratories,

ased on the belief rt rate, pulse ampli- systolic blood pres- erns and the skin's city can be eval- on the computer. ılts now only are erator of the ma- "We hope to de- : errors in human ... bining or which we he margin

controlled ɔnds nega- r, when a e crime is ɔwing the ults from converted polygraph zones, or hich each also as-

Computer Changes Library

A computer is quietly changing tra at the Columbu

What Will Computers Do for You in the Future?

The medical field is an especially attractive one to theorize about. In medicine it sometimes seems that we have too little information (for example, is there any regional trend in the incidence of multiple sclerosis?) and it sometimes seems that we have too much information (notations on minor health problems in thousands of individual files seem more clutter than information). These conflicting aspects suggest that we could make better use of medical information if it were stored better. We might establish a central medical data bank, a computer memory large enough to store all the information, with retrieval possible according to many different criteria. Then a physician called to treat an accident victim, say, could get quickly from the data bank the relevant parts of the victim's medical history, blood type, sensitivity to medications, age, general condition of health, home physician, next of kin, and status of health insurance. Information storage in individual files in doctors' offices has two main defects: The information on some patients may not be complete, and (2) The storage may be in a form that makes retrieval difficult.

Exercise 10–16. Talk to a dentist, an eye doctor, or other specialist about data from the general practitioner's files that he needs for some patients. Is he satisfied with his own data storage on those patients?

Exercise 10–17. Talk to an Emergency Room attendant or a general practitioner about the difficulties of treating a stranger.

However, using computers in management of individual cases is just a beginning. Buried in personal dossiers all over this country and others are statistics of little individual importance that could be valuable if seen as a whole. What are the trends in the incidence of each disorder? What percentage of leukemia patients have a history of serious virus invasion? What are the eating habits, smoking habits, and exercise habits of the population, and what effects have these habits on health? Of course, advances in general medical research ultimately bring benefits to individual patients.

One item of peripheral medical information we mentioned was the status of a patient's health insurance coverage. As a matter of fact, one of the early regional medical data banks has been established by a health insurance company, although nonfinancial information is stored, too (Figure 10–4). This brings us to the use of computers in finance, especially in the world of credit. As science fiction writers postulated long ago, we may eventually have all credit records stored in a central data bank, national, international, or interplanetary, with money itself obsolete. Individual credit cards from issuing companies would also be unnecessary, since just one card or number for each person would provide a rapid report on his available credit. Records could be retrieved as needed for deciding on a new business venture, for preparing income tax forms (perhaps directly by the Internal Revenue Service, rather than by the taxpayer!), and so on.

The management of human population growth in its relation to the total ecology of Earth is now under serious study, and experts on conservation of natural resources warn us that time is running out. Certainly our time is short for real-life, trial-and-error solutions to these problems. It may be possible to simulate some of the problems and suggested solutions on computers instead.

What about the economy? Producers could conceivably have continuous readings of trends in consumption very much like the readings of pulse rate from a patient in intensive care. Consumers also could be provided with wide information on which to base choices.

What about politics and government? At present, an elected representative who tries conscientiously to keep track of opinions in his constituency admits to big gaps in his samples. It would be possible to have votes on issues taken routinely quite frequently, even weekly.

What about education? CAI, Computer Assisted Instruction, is now in lusty infancy. Professor Patrick Suppes* classifies CAI as (1) drill and practice, (2) tutorial, and (3) dialogue. The dialogue programs are especially exciting, as they respond to input from the student, taking him

---

* "On Using Computers to Individualize Instruction," in *The Computer in American Education*, edited by D. D. Bushnell and D. W. Allen. New York, John Wiley & Sons, Inc., 1967.

A COMPUTER TERMINAL.

over easier loops in areas of deficiency, or increasing the pace when he responds correctly.

Many foresee a day when terminals in every household will make available to hundreds of users at once the facilities of huge central computer "utilities." "Timesharing" is a use of a computer utility to execute a number of programs concurrently. Anyone with access to a terminal would have access to the calculating capacity plus all the educational and entertainment programs stored at the central utility. The school student might ask the computer for a square root to use in his homework. The housewife might apply for a short course in football rules during the TV football season, or a list of abundant commodities along with the recipes and menus that use them. The therapist might have his patient study at the terminal the proper use of prescribed medications or practices.

What Are the Drawbacks to Large-Scale Computer Use?

For one thing, a miss is usually a mile. Since computers are often used precisely because the amount of data processing is large, mistakes tend to be big ones. If you have ever been assigned the same gasoline credit card number as a leading commercial airline, or had the decimal point on your water bill moved two digits, you understand the situation.

A severe problem is how to bring the blessings of computers to the individual without squashing him. There are legal problems of how to preserve privacy if anyone with access to central data banks can learn our credit balance, how we plan to vote in the next election, and what our food preferences are. Beyond the threat to privacy, though, is the threat to individual identity. Numbers turn computers on, but they turn many people off. We may have to use CAI to teach people to overcome this prejudice!

Both the error problem and the individual identity problem can be ameliorated if good manual overrides are included with every computerization. For example, the computer could be programmed to alert personnel if it figured a residential utility bill over $200. In fact, increasingly we are learning that computers can free human workers from routine cases, so that they can concentrate on special cases that need them.

Exercise 10–18. Recall some counseling situation with, say, an academic advisor, a guidance counselor, or a medical advisor. Which parts of the interview were routine? Which parts could have been expanded if computers had relieved the routine? Do you feel that the counselor in question "got to know you" during the interview?

Exercise 10–19. Find out what provision is made by a credit-card issuing firm to cover (1) lost credit cards, (2) stolen credit cards, (3) complaints about billing, (4) complaints about the service or product.

An excellent and readable book about computer applications now and in the future is *The Computerized Society* by James Martin and Adrian R. D. Norman (Prentice-Hall, Inc., New Jersey, 1970).

HOW TO COMMUNICATE WITH COMPUTERS

Flowcharting

You are trying to teach someone to eat soup. To simplify it for him, you break the process into a sequence of steps that are already familiar to him, as in Figure 10–6.

FIGURE 10–6. SEQUENTIAL SOUP EATING.

Break the process of eating soup into a sequence of steps. First, pick up the spoon (1), then dip up some soup (2), and taste it (3). Decide whether it is too hot (4). If it is, blow on it (5), and taste it again (3), but if it is not too hot, put it in the mouth (6), and swallow it (7).

Figure 10–7 shows the step-by-step process in schematic form.

The flow chart in Figure 10–7 shows not only the seven steps but also their sequencing. The arrows show the order of the various operations. Different kinds of operations have different-shaped "road" signs:

Start and *Stop* have oval signs:

Operations have rectangular signs:

Decisions (*yes* or *no*) have diamond signs:

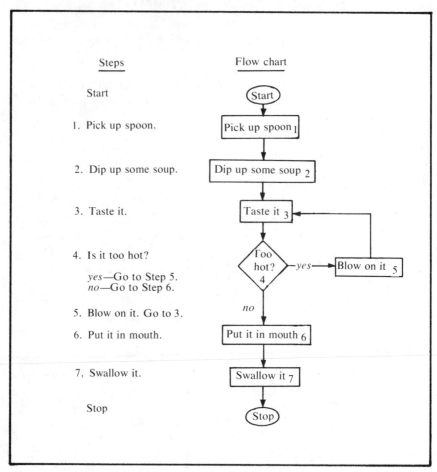

FIGURE 10–7. STEP-BY-STEP DIRECTIONS FOR EATING SOUP.

To make a flow chart for a process, break the process into a sequence of short steps. Use sign shapes to distinguish starts and stops, operations, and decisions. Use arrows to show what follows each step.

Notice that the flow chart in Figure 10–7 contains a loop from

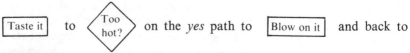

Taste it | to | Too hot? | on the *yes* path to | Blow on it | and back to

Taste it | This is characteristic of directions for a computer. The machine is often told to test results-so-far against a criterion to find out whether to go on or to repeat a subsequence of operations. Some processes "branch" into several routes after a decision.

To make a flow chart for a process, first separate the process into individual operations and decisions. After a start sign, let an arrow show which sign comes next. The *yes* arrow from a decision shows which sign comes next in case the answer is *yes;* the *no* arrow shows which sign comes next in case the answer is *no*. A loop is formed if an arrow leads to an earlier sign in the sequence.

How to take a shower.
First break the process into steps.

> Start
>
> 1. Turn on the hot water.
>
> 2. Wait a while.
>
> 3. Is the water too hot?
>   *yes*—Go to 4.
>   *no*—Go to 5.
>
> 4. Turn on the cold water some.
>   Go to 2.
>
> 5. Step in shower.
>
> Stop

A flow chart for this oversimplified description is shown in Figure 10-8. The loop enclosed in dots calls for retesting after each increase in

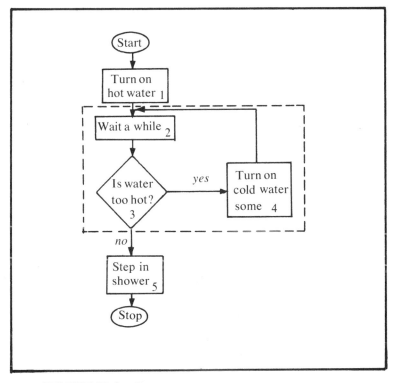

FIGURE 10–8. FIRST FLOW CHART FOR TAKING A SHOWER.

cold water to see whether the temperature is still too hot. As long as it is still too hot (a *yes* at decision 3), follow the *yes* arrow to 4, that is, increase the cold water. What happens when the decision at 3 is *no*? Then the process terminates with a shower.

Admittedly, the flow chart in Figure 10–8 is too simple. It calls for increasing the cold water by small enough steps so that apparently the mixture never becomes too cold. We can improve the flow chart by including a section for testing "Is the water too cold?" We can use the same test loop as before for "Is the water too hot?" and test for "too cold" only if the answer to "too hot?" is *no*. We will also need a new operation box, "Turn cold down some," in case the decision on "too cold?" is *yes*. In Figure 10–9 the improved flow chart is shown, retaining the numbering

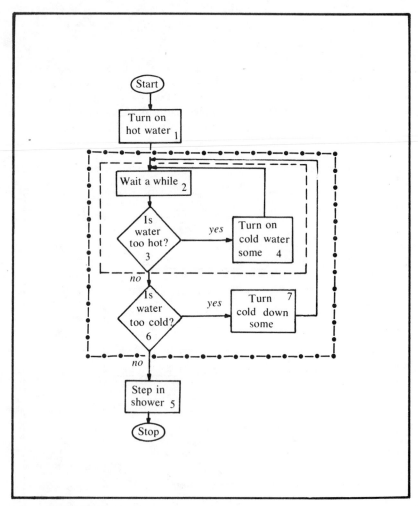

FIGURE 10–9. SECOND FLOW CHART FOR TAKING A SHOWER.

The "too hot" loop (– – – –) and the "too cold" loop (– •–•–) are shown.

of steps from Figure 10–8, even though that makes 5 come after some larger numbers.

Figure 10–10 shows how to use connectors (○) to eliminate some of the visual congestion in this second flow chart. At each *yes* decision the flow chart is broken with a connector labeled *H* (for "too hot") or *C* (for "too cold"). This permits us to treat the "too hot" loop and the "too cold" loop separately, even on a separate page if this one is getting crowded, by dividing the flow chart as shown by the wavy line. At the end of each separate loop is the connector *W* (for "wait a while"), showing where to re-enter the main flow chart.

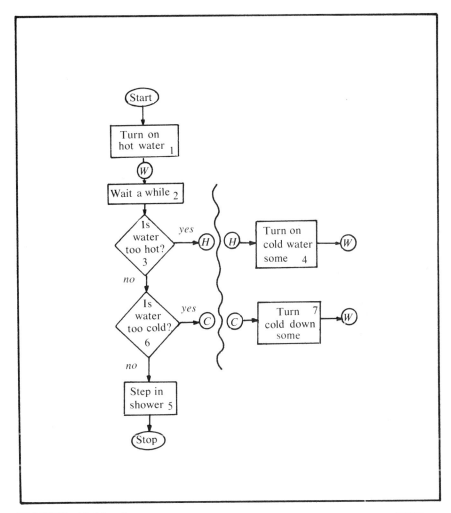

FIGURE 10–10. SECOND FLOW CHART FOR TAKING A SHOWER, WITH CONNECTORS.

In Figure 10–11 we show the regrettable situation of the would-be shower-taker who discovers that there is no hot water available.

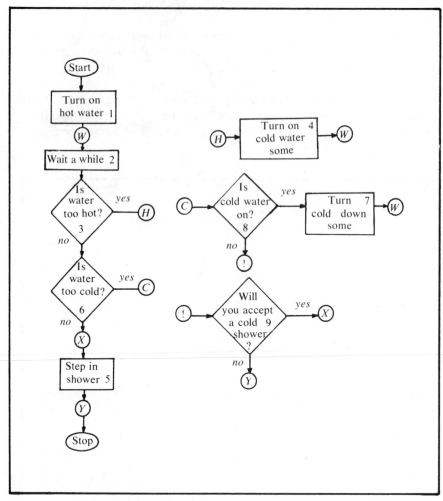

FIGURE 10-11. THIRD FLOW CHART FOR TAKING A SHOWER, WITH THE POSSIBILITY OF NO HOT WATER.

Exercise 10-20. Follow the flow chart in Figure 10-11 enough times to travel at least six possible routes. Describe the process of the shower in each case.

Exercise 10-21. Assign designations to the six connectors, *a, b, c, d, e, f,* in Figure 10-12, show directions by arrows, and label the *yes* arrow and the *no* arrow from each decision diamond, so that the result used with Figure 10-11 handles the case of no cold water. Which part of Figure 10-11 is replaced by Figure 10-12?

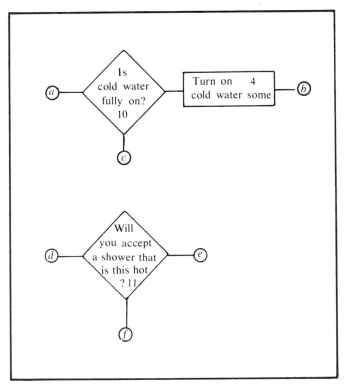

FIGURE 10–12. A NO-COLD-WATER SEQUENCE FOR THE THIRD FLOW CHART.

Exercise 10-22. Design a flow chart for one of these processes:

(a) washing dishes, including scraping, washing, and drying, with instructions in case a dish is presented for drying without being clean

(b) memorizing a list of 20 vocabulary words, including instructions in case you run out of study time before the memorization is complete

(c) arranging a date with someone.

There is quite a lot of latitude in flowcharting. Two charters may make different choices about where to break the chart with connectors, or whether to show the main flow downward or from left to right. There are a few necessary conventions, however. In general the chart should flow either downward or left to right, although loops that are not shown separately with connectors may require some backward flow. Arrows should make it clear which direction the flow takes at each step. From each decision diamond a *yes* arrow and a *no* arrow should show what follows either decision.

The flow chart in Figure 10–13 shows how the payroll might be computed for a business. Each employee, N = 1, 2, . . . , N MAX, where $N\ MAX$ is the number of employees, has an employee number E and an "hours worked" record H. The basic work week in this business

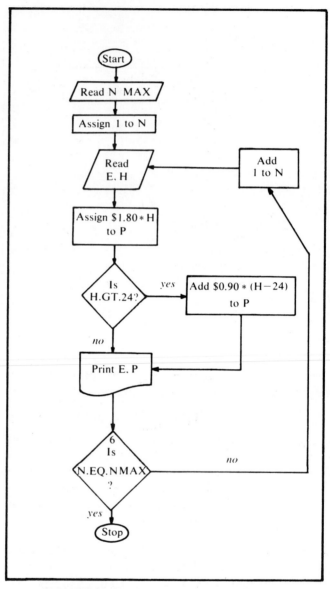

FIGURE 10-13. Flow chart for a payroll.

of the future is 24 hours, and we assume each employee has worked at least that long during the week. For each employee, H gives the hours he has worked. He receives a base pay of $1.80 an hour for the first 24 hours, and if he worked any more than that he qualifies for an extra 90¢ an hour for overtime. His pay P is the total.

Two new sign shapes appear in Figure 10–13, the parallelogram shape, instructing the computer to read something from a table (in this case time-clock records), and the document symbol, instructing the computer to print a result for our inspection.

We have dropped the numbering of steps, since the arrows show the sequencing better than enumeration can.

In the payroll flow chart of Figure 10–13 we use some abbreviations from FORTRAN, a computer language introduced later in this chapter. As in FORTRAN, the star * is used to indicate "times," or "multiplication," the symbol ".GT." means "greater than," ".LT." means "less than," ".EQ." means "equal to." The symbol "N MAX" for "N Maximum" is used as the number of employees. It can be set higher if more are hired or lower if the labor force is decreased.

Exercise 10–23. Sketch an alternative payroll flow chart using connectors.

Exercise 10–24. Make a flow chart for paying salesmen who are paid on commission, $10 for each item costing less than $200, and $15 for each item costing $200 or more. Let S stand for salesman, C stand for commission, N stand for the number of lower-priced items sold by S, and H stand for the number of higher-priced items sold by S.

Exercise 10–25. Make a flow chart for computing the wages of 15 laborers called to light orchard heaters in orange groves during a freeze. The laborers are paid $10 each time they are called out. If they work longer than five hours, they are paid at the rate of $2 an hour for each additional hour.

Figure 10–14 shows a flow chart for multiplying two 2 by 2 matrices (see Chapter 5). In this case we are going to use the product matrix in further calculations, but we are not interested in it for itself. Therefore, we do not have the computer print it (document symbol), but have it filed on magnetic tape, with the new symbol Q .

Exercise 10–26. Compare the flow chart of Figure 10–14 with any explanation in ordinary English of the multiplication of matrices. (You may use the explanation in Chapter 5, or any other explanation.) Which features are clearer in the flow chart, and which are not so clear?

Exercise 10–27. Show what to change in Figure 10–14 to cover multiplication of 3 by 3 matrices. It is possible to find the product AB of an r by n matrix A and an n by c matrix B. Show that by using r, c, and n as maxima for i, j, and k, respectively, you can adapt the flow chart for this purpose.

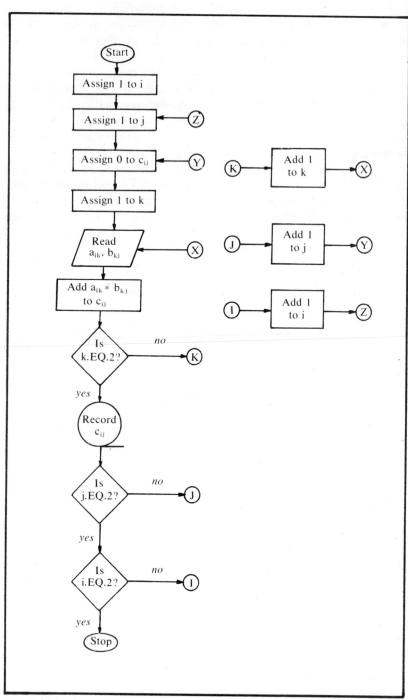

FIGURE 10–14. Flow chart for multiplying two 2 by 2 matrices, A and B, to find C = AB.

The flow chart in Figure 10–15 covers the matching process described in the dating-service process in Chapter 1, page 3. The chart looks rather complicated, but each step is simple, so let us trace the process step by step with an example. In Chapter 1 each boy and each girl had a name, but here we use numbers instead to adapt to the computer's capabilities. In Chapter 1 we were given a square array with two numbers in each row-column position. Here we separate this information into two arrays, a boys' preference "matrix" (here meaning simply "rectangular array") P and a girls' preference matrix Q.

P	g_1	g_2	g_3	g_4		Q	g_1	g_2	g_3	g_4
b_1	1	2	4	3		b_1	2	1	2	1
b_2	3	1	4	2		b_2	3	3	3	3
b_3	2	1	4	3		b_3	1	2	4	2
b_4	1	2	4	3		b_4	4	4	1	4

From the second row of the boys' preference matrix P we can tell that boy 2 likes girl 2 best, girl 4 second-best, then girl 1, and, last, girl 3. Each row consists of the numbers 1, 2, 3, 4 in some order, depending on how the boy ranked the four girls. From the girls' preference matrix Q we can read from the third column that girl 3 likes boy *4* best, then boy *1*, then boy *2*, and, last, boy *3*. Each column consists of the numbers *1*, *2*, *3*, *4* in some order, depending on how the girl ranked the four boys.

The object of the dating-service process is to produce a matching that dates each boy with exactly one of the girls, following preferences as much as possible. The solution will be expressed in a "Matching" matrix M of 0's and 1's, a 0 in row b, column g if the boy and girl are not matched, a 1 if they are matched. Then each boy's row should have exactly one 1 in it, the rest 0's, and each girl's column should have exactly one 1 in it, the rest 0's. Since the computer will be asked to update the trial matches in stages as we go along, we give it an initial "solution" matrix M of all zeros to reserve the necessary storage space in the computer memory, erasing old storage from other problems, just as newly recorded material erases what it replaces on a tape recording. The initial matrix M also contains a dummy row of zeros called row 0, because it will simplify notation later. We will explain that when we get to it, but for now it suffices to say that the 0th row will always have only 0's in it.

M

b	g	1	2	3	4
	0	0	0	0	0
	1	0	0	0	0
b	2	0	0	0	0
	3	0	0	0	0
	4	0	0	0	0

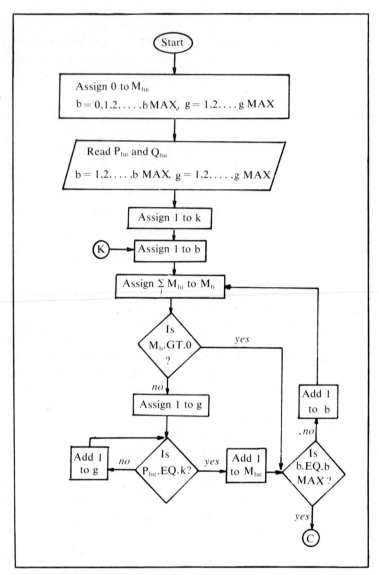

FIGURE 10–15. FLOW CHART FOR DATING SERVICE PROCESS.

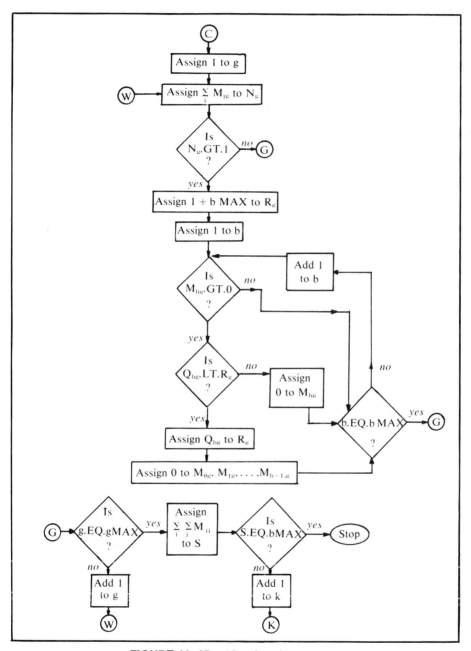

FIGURE 10–15 (Continued).

We have covered the first three instructions of Figure 10–15, "Start," "Assign all 0's to M," and "Read the two preference matrices." At the next stage we "Assign 1 to k." Here "k" stands for the boys' ranking index. In this first stage the process matches each boy to his first (rank $k = 1$) choice. Then we "Assign 1 to b," that is, consider how to match boy 1 first. The next step calls for finding $\sum_j M_{bj}$, the sum of all entries in boy b's (here boy 1's) row, and calling the sum M_b, here M_1. Since initially the matching matrix M_{bg} is all 0's, the sum of boy 1's row is zero, and $M_1 = 0$.

The next step is a "decision." "Is M_b greater than 0?" As you will see at a later stage, we are asking, "Does this boy already have a date?" However, since no matches have been made yet, the answer is "*no*," which carries us to the step "Assign 1 to g," or simply "Consider the first girl." Next the decision "Is P_{11} equal to 1?" The computer looks at the preference values we had it read to determine whether girl 1 happens to be boy 1's first choice. Since the answer is "*yes*," we "Add 1 to M_{11}." "Is b equal to the maximum b, which in this case is 4?" *No.* Then "Add 1 to b," that is, go on to the next boy. Next sum row 2 of the Matching matrix M_{bg}, calling the result M_2. The Matching matrix now has one 1 in it, but that is in row 1. Row 2 is entirely made up of 0's, so M_2 is 0.

This time when the computer looks at row 2, column 1 of P_{bg}, it finds a "3," so the answer to the decision "Is P_{21} equal to 1" is "*no*," whereupon we add 1 to g and try again. *Yes*, b_2's first choice is girl 2, so $P_{22} = 1$. Then "Add 1 to M_{22}" in the Matching matrix.

Exercise 10–28. Show that after every boy has been matched to his first choice the Matching matrix looks like this:

M

		g			
		1	2	3	4
	0	0	0	0	0
	1	1	0	0	0
b	2	0	1	0	0
	3	0	1	0	0
	4	1	0	0	0

Exercise 10–29. Find the step on the control chart at which all four boys have been given their first choices. Where does the chart direct us next?

One great advantage you have over a computer is that you are able to scan several figures at once and tell at a glance which is largest, or whether one of them is a 1. The computer has to check them one by one. The computer does not simply "match each boy to his rank 1 choice."

It takes one boy at a time and then looks at one girl at a time, asking, "Is this girl his first choice? Is this one?"

So far our Matching matrix is not one we can use, for two of the girls have two dates and two have no dates. Our rule for resolving a "conflict" of more than one date for a girl is to have the girl reject all but the applicant she likes best. How does the flow chart accomplish this? Starting at "C" (for "conflict"), "Assign 1 to g," that is, start with girl 1. Find $\sum_{i=1}^{bMAX} M_{i1}$, the sum of column 1 of the Matching matrix so far, and call the sum N_1. N_1 tells how many dates girl 1 has so far, because there is a 1 for each date and a 0 for each boy she is not dating.

"Is N_1 greater than 1?" *Yes.* There is a conflict in her case. N_1 is 2. "Assign the value $1 + b$ MAX (here 5) to R_1." So far this is an artificial index, but in a few steps it will be erased by newer information about girl 1's rankings of the boys who asked to date her. Once again, the computer does not simply scan the Matching matrix and notice that it is boys 1 and 4 who are matched to her. The computer looks at each boy in turn to see whether he is matched to her, by asking, "Is M_{b1} greater than 0 (hence 1) for this boy b? for this one? The answer for boy 1 is "yes," so go on to the question "Is Q_{11} less than $R_1 = 5$?" Yes, all the ranks are less than 5, since they are *1, 2, 3, 4*. Then assign 2 to R_1, replacing the artificial 5. Assign 0 to M_{01}, that is, to all the Matching matrix entries in girl 1's column *before* this boy 1 who is dated to her. In this case only the entry in the dummy row is affected and it remains zero. The dummy row was included so that we could talk about boys "previous" to the first.

"Is 1 equal to b MAX?" No, b MAX is 4. Then "Add 1 to b," that is, go on to the next boy. Is he matched to girl 1? *No.* Is 2 less than 4? *Yes.* Add 1 to 2. Is boy 3 matched to girl 1? *No.* Is 3 less than 4? *Yes.* Add 1 to 3. Is boy 4 matched to girl 1? *Yes.* Now have the computer check how girl 1 ranked boy 4. Q_{41} is *4*, showing that she liked him least. How does this ranking, 4, compare with the rank R_1 of 2 that we already have stored for girl 1 because of her date with boy 1? Is 4 less than 2? No, so "Assign 0 to M_{41}." This cancels her date with boy 4, since she prefers boy 1 to him. "Is 4 equal to b MAX?" *Yes,* so go to connector "*G.*"

Exercise 10–30. From connector G go to the decision "Is 1 equal to g MAX," answer "*no*," and follow connector W back into the "conflict" chart. Find N_2, and show that there is a conflict over girl 2.

Which boys are currently dated to girl 2? First the computer tries boy 1. $M_{12} = 0$, so boy 1 is not dated to her. $M_{22} = 1$, so boy 2 is dated to girl 2. So far R_2 has the value $1 + b$ MAX $= 5$. How does girl 2 rank boy 2? $Q_{22} = 3$, which is less than 5, so replace 5 by 3 as value for R_2. Now try boy 3. Yes, $M_{32} = 1$, so boy 3 is also dated to girl 2. How does she rank him? $Q_{32} = 2$, which is less than the R_2 of 3 that we have stored because of her ranking of boy 2. Then reassign the R_2 value to be 2, cancel her date with boy 2. ("Assign 0 to M_{02}, M_{12}, M_{22}." The only real change is in M_{22}, which had been 1.)

Exercise 10-31. Follow the flow chart, showing that boy 4 is not dated to girl 2. Go to connector G, add 1 to g, and show that there is no conflict over girl 3. Then show that there is no conflict over girl 4.

At this point, after Exercise 10-31, the answer to "Is g equal to g MAX?" is "*yes*." "Assign $\sum_i \sum_j M_{ij}$ to S." This amounts to counting the matches that still remain after all conflicts have been decided. In this case S is 2, for girls 1 and 2 have dates, but girls 3 and 4 have not. Then add 1 to k, making it 2, and return to the first part of the flow chart to reassign rejected boys to their second choices.

Exercise 10-32. Follow the K chart with k = 2, showing that after boys rejected in the first round are given their second choices the Matching matrix is

M

$$
\begin{array}{c|cccc}
 & \multicolumn{4}{c}{g} \\
 & 1 & 2 & 3 & 4 \\
\hline
0 & 0 & 0 & 0 & 0 \\
\cdots & \multicolumn{4}{c}{\cdots\cdots\cdots\cdots} \\
1 & 1 & 0 & 0 & 0 \\
b \quad 2 & 0 & 0 & 0 & 1 \\
3 & 0 & 1 & 0 & 0 \\
4 & 0 & 1 & 0 & 0 \\
\end{array}
$$

Exercise 10-33. Go on to the conflict chart. Show that there is no conflict over girl 1, but that there is a conflict over girl 2. Show that she rejects boy 4.

Exercise 10-34. There are no conflicts over girls 3 or 4. Find S = 3. Which girl has no date? Re-enter at connector K, and assign the one rejected boy to his third choice.

Exercise 10-35. Complete the dating assignment with S = 4.

We have used b MAX and g MAX instead of 4, so that we can generalize readily to a larger example.

Exercise 10-36. Find loops in Figure 10-15 that show where the computer is considering one boy at a time. Find loops for one girl at a time.

This flow chart shows the typical feature of loops followed around and around with a decision in each cycle "Are there still more cases to consider?" Also, note the branching at each decision, a *yes* answer taking the process into one path, a *no* answer into another.

You can find more examples of flow charts and exercises with solutions in *Flowcharting* by Mario V. Farina (Prentice-Hall, Inc. New Jersey, 1970.)

Programming

Flow charts help clarify how a process works, how it loops back on itself, how it branches, and where decisions are called for. Now we need a way to communicate the charted process to the computer. What we want is a way to present our general needs, much as we might ask a person to "come over here," without describing in physiological detail what messages his nerves should take to which muscles to accomplish the first step, then the second, and so on. How do computers count, add, multiply? The answer depends on the machine. A computer may do arithmetic by switches, vacuum tubes, transistors, magnetic tape, thin films, circuits, cores, or combinations of these and other components. For that matter, how closely can we describe how a human being adds numbers or multiplies them? How does he store or retrieve information in his memory?

Several "compiler" languages have been developed to translate instructions from a human programmer in a language oriented to his needs into the basic machine language, which uses the switches, transistors, or other components.

Some well known compiler languages are

COBOL, especially for business-oriented problems,

ALGOL, an "algorithm-oriented" program (an algorithm is a step-wise procedure such as long division or the dating-service process of Chapter 1), and

FORTRAN, one of the first compiler languages.

Another compiler language, **PL/I** (Programming Language I), contains elements from COBOL, ALGOL, and FORTRAN to facilitate scientific and business interchange. This compiler language, which explains a programmer's problem-oriented instructions to the computer, has a *super*language, called a "preprocessor," NEATER 2,* which reviews a PL/I program, improving the format and pointing up clerical errors and some of the logical errors the programmer has made.

As computers are improved, so are the computer languages and the compiler languages. There are quite a few compiler languages in existence now, most of them undergoing revision continually. Each has its proponents and a set of problems for which it is efficient. Here we introduce some important elementary parts of FORTRAN IV. "FORTRAN" is a contraction of the words "formula translation." The programmer writes in a language reminiscent of mathematical formulas. Then the language compiler translates this to the basic machine language of the particular computer he is using.

---

* Developed by Kenneth Conrow and Ronald G. Smith, "Communications of the Association for Computing Machinery," Vol. 13, No. 11, November, 1970, pages 669–675.

CONSTANTS AND VARIABLES

We are going to use the computer on formulas involving constants and variables. In FORTRAN a **constant** is a numerical value. There are two kinds of constants that FORTRAN recognizes—**integer constants** and **real constants.** An integer constant is always a whole number, positive, negative, or zero, and *never* has a decimal point, while a real constant *always* has a decimal point.

In the statements

$$N = 20 + I$$
$$R = 20.0 + S$$

the constants are 20 and 20.0. Of these, 20 is an integer constant and 20.0 is a real constant, and while to us they are essentially the same, they are very different to the machine. (Some computers are advanced enough to convert all input to the same mode, if the programmer includes both integer and real constants, but others must be programmed entirely in one mode.)

Integer Constants	Real Constants
0	0.0
1	17.6
−200	.00043
11764	−201.613

Notice that commas are not used to separate triads of digits in the numeral 11764. A comma has a technical interpretation in FORTRAN.

EXAMPLE   In Figure 10–16 we will show a FORTRAN program for computing a payroll. Although the 24 hours used to determine overtime is an integral number of hours, it is written "24.0" in the program, to be consistent with the real mode.

The term **variable** is used to denote any quantity that is referred to by name, such as X, I, $GEORGE$, $I17J$. A variable may have many different values during any given program. In the statement

$$GEORGE = 3.1416 - X + PSI,$$

$GEORGE$, X, and PSI are variables, while 3.1416 is a constant.

Since FORTRAN has different arithmetics when using integers and reals, it is necessary to avoid mixing the real-valued objects and the integer-valued objects in the same FORTRAN statement. Therefore, it is necessary to have a convention that will tell the machine which variables take on integer values and which take on real values. If the name of a variable starts with I, J, K, L, M, N, it is an integer-valued variable. If it starts with *any other letter*, it is a real-valued variable. The first symbol in a variable name must be a letter, but any of the rest of the up-to-six letters a name may have can be either a letter or a number. Special symbols, such as parentheses, comma, period, star, and so on, may not be used in a name.

In the following list the real or integer character of the variable and the constant value assigned to it agree.

$$J = 173$$
$$RHO = -0.0076$$
$$GEORGE = 1.0$$
$$LBJ = 36$$
$$RMN = 37.0$$
$$J37T = -72$$

Exercise 10–37. Tell which of the following have real values, and which integer values:

(a) 1.0   (e) HENRY   (i) −2
(b) 17   (f) −14.7   (j) X
(c) I12   (g) 7711654   (k) J
(d) IRMA   (h) 1766532.0   (l) YOU

Exercise 10–38. Which of the following are illegal and why?

(a) I + 7   (d) GEORGE + HENRY   (g) IXY = 17
(b) M + 14.6   (e) HOMER + IRMA   (h) J12 = 13.6
(c) X + 22   (f) 7 + 8.0   (i) 3KL = 4

Exercise 10–39. Select names that might be assigned the following constants, consistent with the conventions for integer- and real-valued names.

(a) −100.2   (d) −176.0   (g) −0.003
(b) 0   (e) 28   (h) 0.0
(c) 14.6   (f) 11176432   (i) 116.732

ARITHMETIC INSTRUCTIONS

The basic arithmetic instructions used in FORTRAN are the following:

Name	Symbol	Example	Meaning
addition	+	C = A + B	The values assigned to *A* and *B* are added and the sum is assigned to the variable *C*.
subtraction	−	D = E − F	The value assigned to *F* is subtracted from the value assigned to

			E and the difference assigned to the variable D.
multiplication	*	GEORGE = X * SUE	The value assigned to X is multiplied by the value assigned to SUE and the product is assigned to the variable $GEORGE$.
division	/	QUO = R/S	The value assigned to R is divided by that assigned to S and the quotient assigned to the variable QUO.
exponentiation	**	POW = R ** 2.0	The value assigned to R is raised to the power 2.0 and the result assigned to the variable POW, so that $POW = R^2$.

At this point we must mention that it is permissible to raise a real-valued constant or variable to an integer-valued exponent, so POW = R ** 2 is also legal.

FORTRAN includes a specified order for performing several operations in the same statement: exponentiation is performed first, multiplication and division next, and addition and subtraction last. Parentheses can be used to group operations to form a single term and should be used if there is any doubt as to the order in which operations will be performed. For example,

$$A = B/D + E$$

means

$$A = \frac{B}{D} + E,$$

and

$$A = B/(D + E)$$

means

$$A = \frac{B}{D + E} .$$

$$A = B * D + E$$

means

$$A = (BD) + E,$$

while

$$A = B * (D + E)$$

means

$$A = B(D + E) = BD + BE.$$

In FORTRAN two operational symbols cannot appear side by side, but must be separated by parentheses. For example,

$$A/-Y \text{ is not legal;}$$

use

$$A/(-Y).$$

$$A + -B \text{ is not legal;}$$

use

$$A + (-B).$$

$X = U - V + C$ means $x = u - v + c$, while
$X = U - (V + C)$ means $x = u - (v + c)$.

$X = U * V/Y * Z$ means $x = u \cdot \dfrac{v}{y} \cdot z$, because $*$ and $/$ are

taken in order, left to right, while

$X = U * V/(Y * Z)$ means $x = \dfrac{u \cdot v}{y \cdot z}$.

$X = A * B ** 3$ means $x = a \cdot b^3$, while
$X = (A * B) ** 3$ means $x = (a \cdot b)^3$.

Exercise 10–40. Write FORTRAN statements for each of the following:

(a) $a = bc^2 + d$ 　　(d) $a = \dfrac{b}{(cd)^2}$

(b) $a = \dfrac{bc^2}{d}$ 　　(e) $a = b(c + d)^2$

(c) $a = \dfrac{b}{c^2 d}$ 　　(f) $a = b(c + d^2)$

Exercise 10–41. Write FORTRAN statements for the following:

(a) $x = \dfrac{3 \cdot b + \dfrac{a + b}{c + d}}{\left(\dfrac{a + b}{c + d}\right)^2}$ 　　(b) $x = \left[\dfrac{1 + \dfrac{a + b}{c + d}}{a + b + c + d}\right]^2 \cdot \dfrac{a}{b}$

Exercise 10–42. Write mathematical equations equivalent to the following FORTRAN statements:

(a) $\text{AREA} = B * H/2.0$

(b) $\text{VOL} = 4.0 * 3.1416 * R ** 3/3.0$

(c) $S = ((X + Y) ** 3/((X - Y)/Z) ** 0.5) ** 2$

EXAMPLE   The FORTRAN program shown in Figure 10–16 contains the instruction

$$P = P + 0.90 * (H - 24.0),$$

where H is a number of hours worked. This instruction tells the computer to subtract 24.0 from the number of hours worked, to find the hours of overtime, $(H - 24.0)$. Then multiply by 0.90, the "overtime differential" in pay, to find the overtime pay earned. Then add the overtime pay to the base pay P to compute the worker's total pay.

Exercise 10–43. If B = 3.0, C = −1.0, and D = 2.0, evaluate the FORTRAN statement for A:

    (a) A = B * (C + D ** 2)

    (b) A = B * (C + D) ** 2

    (c) A = B/(C * D ** 2)

Exercise 10–44. If I = 3, J = 2, and K = −4, evaluate M in the following FORTRAN statements:

    (a) M = I * J/K

    (b) M = I * K + J/I

    (c) M = K ** 2/(I * J)

LOGICAL STATEMENTS

High-speed computers have the built-in capability to perform the logical operations $\wedge$, $\vee$, and $\sim$, but since these are not compatible with the usual integral- or real-valued constants or variables, logical variable names must be explicitly stated.

A statement **LOGICAL B, C** specifies that B and C are logical variables, that is, take on only the values True and False.

The logical operations are .NOT., .AND., .OR., and the periods are part of the operational symbol.

Expression	Value of Expression
.NOT.B	True if B is False; False if B is True
B.AND.C	True if both B and C are True; False if either B or C is False
B.OR.C	True if either B or C is True; False if both B and C are False

TRANSFER STATEMENTS

Several of the flow charts you have studied include decisions capable of moving the flow on to a new part of the process. For instance, in Figure

10–13 the flow goes repeatedly around the loop after the *no* arrow at decision 6 "Is E equal to E MAX?" until there is a *yes* decision; that is, until E equals E MAX. At that point the flow is transferred to a different part of the process. In programs these important changes are accomplished by **transfer statements.** These are of three types: unconditional transfers, arithmetic transfers, and logical transfers.

The **unconditional transfer** is

GO TO n,

which sends the control to the statement named *n*. The other two kinds of transfers are called conditional transfers, because transfer of control takes place only if certain conditions are met. These transfers are in the form of IF statements. A logical transfer is activated if a logical quantity is True and not if it is False. An arithmetic transfer depends on whether an arithmetic expression is negative, zero, or positive.

The **arithmetic transfer** statement has the form

IF (a) n_1, n_2, n_3

The expression *a* must be arithmetic and have a definite value at the place in the program where the instruction is used. If *a* is negative, control is shifted to the instruction named n_1. If *a* is 0, the control is shifted to the instruction named n_2. If *a* is positive, control is shifted to the instruction named n_3.

"IF (X) 10, 20, 30" means: If *x* is negative, go to statement 10; if *x* is zero, go to statement 20; if *x* is positive, go to statement 30. The flow chart in Figure 10–17 calls for just this sort of 3-way decision. In the corresponding FORTRAN program, Figure 10–18, the arithmetic transfer statement appears in the fifth line. **EXAMPLE**

A **logical transfer** statement has the form

IF (*a*) STATEMENT

where *a* is a logical expression (has value True or False) and STATEMENT is an arithmetic statement or a GO TO statement. The logical transfers are used with the six relational operators:

Symbol	Math Symbol	Use	Meaning
.GT.	$>$	A.GT.B	*a* is greater than *b*
.GE.	$\geq$	A.GE.B	*a* is greater than or equal to *b*
.EQ.	$=$	A.EQ.B	*a* is equal to *b*
.NE.	$\neq$	A.NE.B	*a* is not equal to *b*
.LE.	$\leq$	A.LE.B	*a* is less than or equal to *b*
.LT.	$<$	A.LT.B	*a* is less than *b*

EXAMPLE

IF (A.GT.B) X = 10 − Y means

If A is greater than B, then X = 10 − Y; otherwise, go to the next step in the program.

IF (A.EQ.B) GO TO 20 means

If A equals B, transfer control to step 20. Otherwise (that is, if A is not equal to B), go to the next step in the program.

EXAMPLE

Line five of the FORTRAN payroll program in Figure 10–16 shows a logical transfer statement,

$$\text{IF (H.GT.24.0) GO TO 10}$$

If H, the hours worked, is greater than 24.0, then the employee is entitled to overtime pay, which is computed in step 10 and added to his base pay. If he did not work more than 24.0 hours, then the computer advances to the next line, which assigns him base pay only.

OTHER INSTRUCTIONS

READ, PRINT, STOP, and END instructions are also used in programs. Every program must end with the instruction END. Often programs include a STOP instruction after a section that needs checking, or the machine may be programmed to stop in case it gets a negative answer to a problem that should not have one, for instance. One purpose of including STOP instructions is to facilitate debugging (finding mistakes). READ and PRINT instructions are used to get information into and out of the machine. Their proper use varies a great deal from machine to machine, and we shall not go into details, but use one form here. We use

$$\text{READ, A, B, GEORGE, X, Y}$$

to make the machine read values called A, B, $GEORGE$, X and Y into its memory. If we want the machine to read and store in its memory a list of n values

$$I(1), I(2), \ldots, I(n),$$

we say

$$\text{READ, I(k), k = 1, 2, 3, \ldots, n.}$$

If we want it to read and store a two-dimensional array, like

$$I(1, 1), I(1, 2), \ldots, I(1, m)$$
$$I(2, 1), I(2, 2), \ldots, I(2, m)$$
$$\ldots\ldots\ldots\ldots\ldots\ldots\ldots$$
$$I(n, 1), I(n, 2), \ldots, I(n, m)$$

we say

$$\text{READ, } I(i, j), i = 1, 2, \ldots, n$$
$$j = 1, 2, \ldots, m.$$

Similar conventions for PRINT statements enable us to tell the machine to record a result on magnetic tape, punched cards, or its low-speed memory (such as a disk file), or to print a result on paper.

While there are other instructions available—some of them very powerful and sophisticated—you have enough here to start writing FORTRAN programs.

At first, writing programs can be frustrating, as mistakes, clerical and logical, keep intruding. The computing machine does exactly what it is told, and while it may be perfectly clear to any idiot what you intended to say, the machine is not up to idiot level and does what you told it to do. There are ancient folk tales and modern children's stories based on the hilarious incongruities that ensue when someone takes instructions too literally. It is a shock to discover how hard it is to say precisely what we mean.

Figure 10–16 shows a FORTRAN program for the payroll flow chart of Figure 10–13.

```
READ, N MAX
N = 1.0
5 READ, E, H
P = 1.80 * H
IF (H.GT.24.0) GO TO 10
GO TO 20
10 P = P + 0.90 * (H − 24.0)
20 PRINT, E, P
IF (N.EQ.N MAX) GO TO 30
N = N + 1.0
GO TO 5
30 STOP
END
```

FIGURE 10–16. FORTRAN PROGRAM FOR PAYROLL FLOW CHART.

In this program real values are used, so the variable names N, E, H, and P are chosen accordingly, and even values that are initially integers, like the count N of employees, are written with decimal points as real numbers. At the step numbered "5" the computer reads the employee's number and hour record for the week, probably from his punched time card.

Exercise 10–45. Play the role of the computer. Follow the program of Figure 10–16, taking *E MAX* to be 5, with the following information available to you from punched time cards:

N	Employee Number	Hours Worked
1	40312	24
2	3827	24
3	11133	34
4	441	39
5	5459	24

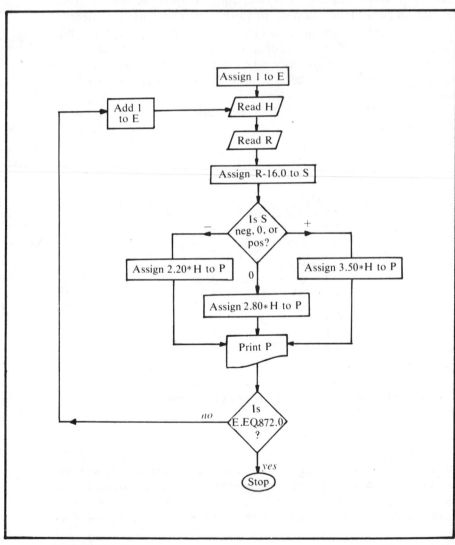

FIGURE 10–17. FLOW CHART FOR PAYROLL COMPUTATION OF 872 EMPLOYEES IN THREE SHIFTS.

Here is another example involving a payroll, this time for 872 employees working in three shifts at different pay scales, as follows:

Shift	Reporting Time (24 – hour clock)	Pay Rate Per Hour
Day	8:00	$2.20
Swing	16:00	2.80
Graveyard	24:00	3.50

Our first step might be to draw a flow chart for the payroll calculation, as in Figure 10–17. Let E stand for Employee, let H stand for Hours, and let R stand for Reporting Time. In this example the employees are simply numbered from $E = 1$ to $E = 872$, so we do not put in a counting index N.

The first decision calls for a three-way transfer statement, as shown in the FORTRAN program in Figure 10–18.

```
      E = 1.0
  5 READ, H
    READ, R
    S = R − 16.0
    IF (S) 10, 20, 30
 10 P = 2.20 * H
    GO TO 40
 20 P = 2.80 * H
    GO TO 40
 30 P = 3.50 * H
 40 PRINT, P
    IF (E.EQ.872.0) GO TO 50
    E = E + 1.0
    GO TO 5
 50 STOP
    END
```

FIGURE 10–18. FORTRAN PROGRAM FOR PAYROLL COMPUTATION FOR 872 EMPLOYEES IN THREE SHIFTS.

Notice that FORTRAN instructions are numbered only if we need to refer to them in a "GO TO" step, and that the numbers chosen are quite arbitrary. As the programmer writes the GO TO step, he inserts a number far enough away to allow scope for earlier steps that may intervene.

Figure 10–14 shows a flow chart for multiplying two 2 by 2 matrices to get $C = AB$. Figure 10–19 shows a FORTRAN program for transmitting the process to a computer. The integral mode was chosen this time, since we are supposing that the two matrices have integral entries, and all other numbers are integers. The matrices A, B, C are given integral names LA, MB, KC.

```
      I = 1
   7 J = 1
   8 KC(I,J) = 0
      K = 1
   9 READ, LA(I,K)
      READ, MB(K,J)
      KC(I,J) = KC(I,J) + LA(I,K) * MB(K,J)
      IF (K.EQ.2) GO TO 10
      K = K + 1
      GO TO 9
  10 PRINT, KC(I,J)
      IF (J.EQ.2) GO TO 20
      J = J + 1
      GO TO 8
  20 IF (I.EQ.2) GO TO 30
      I = I + 1
      GO TO 7
  30 STOP
      END
```

FIGURE 10-19. FORTRAN PROGRAM FOR FINDING C=AB, WHERE A AND B ARE 2 BY 2 MATRICES.

The following exercises give you a chance to write some FORTRAN programs yourself, following the examples you have seen.

Exercise 10-46. Write a FORTRAN program for totaling students' grades. Students receive 10 points for each question answered correctly. Those who also worked an extra credit question receive 15 bonus points. There are 1355 students who took this examination. Check your program for these features:

1. Did you use integral mode?
   a. If yes, did you use corresponding integer names?
   b. If no, did you write all real· numbers correctly, with decimal points?
2. Did you have the computer read each item of information you had it use?
3. In what form will you use the answers? Did you have the computer print or store them?
4. Did you have the computer check each time to see whether it had finished with the 1355th student? What should it do if it has not? What should it do if it has?

Exercise 10-47. Write a FORTRAN program for totaling the admissions paid for a concert. Of 970 tickets taken at the door, some are yellow tickets sold to adults at $4, and some are childrens' red tickets sold at $2.

Exercise 10-48. Suppose in Exercise 10-47 that some of the 970 tickets were free passes. Write a FORTRAN program to take care of all three ticket prices: $4, $2, and $0.

```
FORTRAN IV G LEVEL  18              MAIN           DATE = 72090        12/59/11        PAGE 0001

C001          CIMENSION LA(2,2),MB(2,2), KC(2,2)
C002          I=1
C003        7 J=1
C004        8 KC(I,J)=0
C005          K=1
C006        9 READ(5,101) LA(I,K),MB(K,J)
C007      101 FORMAT(2I2)
C008          KC(I,J)=KC(I,J)+LA(I,K)*MB(K,J)
C009          IF(K.EQ.2) GO TO 10
C010          K=K+1
C011          GO TO 9
C012       1C WRITE(6,102) KC(I,J)
C013      102 FORMAT(I10)
C014          IF(J.EQ.2) GO TO 20
C015          J=J+1
C016          GO TO 8
C017       2C IF(I.EQ.2) GO TO 30
C018          I=I+1
C019          GO TO 7
C020       3C CONTINUE
C021          END
```

FIGURE 10-19 (a)

This shows the FORTRAN IV program of Figure 10–19 as actually run on a computer. The computer listed the execution time— .002 hrs, or 7.2 seconds!

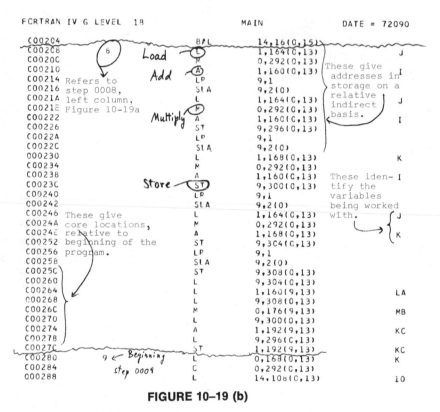

FCRTRAN IV G LEVEL 18   MAIN   DATE = 72090

C00204			BPL	14,16(0,15)	
C00208		Load	L	1,164(0,13)	J
C0020C			M	0,292(0,13)	
C00210		Add	A	1,160(0,13)	These give
C00214	Refers to		LP	9,1	addresses in I
C00216	step 0008,		SlA	9,2(0)	storage on a
C0021A	left column,		L	1,164(C,13)	relative
C0021E	Figure 10-19a		M	0,292(0,13)	indirect J
C00222		Multiply	A	1,160(C,13)	basis.
C00226			ST	9,296(0,13)	I
C0022A			LP	9,1	
C0022C			SlA	9,2(0)	
000230			L	1,168(C,13)	K
C00234			M	0,292(0,13)	
C00238			A	1,160(C,13)	These iden- I
C0023C		Store	ST	9,300(0,13)	tify the
C00240			LP	9,1	variables
C00242			SlA	9,2(0)	being worked
C00246	These give		L	1,164(C,13)	with. J
C0024A	core locations,		M	0,292(0,13)	
C0024E	relative to		A	1,168(0,13)	K
C00252	beginning of the		ST	9,3C4(C,13)	
C00256	program.		LP	9,1	
C00258			SlA	9,2(0)	
C0025C			ST	9,308(0,13)	
C00260			L	9,304(C,13)	
C00264			L	1,160(9,13)	LA
C00268			L	9,308(0,13)	
C0026C			M	0,176(9,13)	MB
C00270			L	9,300(0,13)	
C00274			A	1,192(9,13)	KC
C00278			L	9,296(C,13)	
C0027C			ST	1,192(9,13)	KC
C00280	9 ← Beginning		L	0,168(0,13)	K
C00284	step 0009		C	0,292(C,13)	
000288			L	14,108(C,13)	10

FIGURE 10–19 (b)

This shows one step of the FORTRAN IV program of Figure 10–19(a) expanded into the basic machine language in the computer. The compiler language enables the human programmer to avoid the work of talking this complicated language.

PL/C dialect of PL/I

```
*PL   XREF,ATR,PAGES=10,TIME=(0,30)

*OPTIONS IN EFFECT*   PAGES=010,TIME=(000,030),ERRORS=(020,015),SORMGIN=(002,072,001),LINECNT=058,
*OPTIONS IN EFFECT*   FLAGW,XREF,ATR,NOLIST,UDEF,SOURCE,NODUMP,NODUMPARRAY,NOM91,CHECK

MULT:PROCEDURE OPTIONS(MAIN);                          PL/C-R5.4562  03/30/72 13:27 PAGE  1

STMT LEVEL NEST BLOCK   SOURCE STATEMENT                                         ID FIELD

  1              MULT:PROCEDURE OPTIONS(MAIN);
  2    1    1         DCL (R,C,N) FIXED DECIMAL;
  3    1    1         GET LIST(R,C,N);
  4    1    1         BEGIN;
  5    2    2         DCL (A(R,C),B(C,N),D(R,N)) FIXED DECIMAL;
  6    2    2         I=-1;
  7    2    2    SEVEN:J=1;
  8    2    2    EIGHT:D(I,J)=0;
  9    2    2    K=1;
 10    2    2    NINE:GET LIST(A(I,K),B(K,J));
 11    2    2         D(I,J)=D(I,J)+A(I,K)*B(K,J);
 12    2    2         IF K=N THEN DO; PUT LIST(D(I,J));
 15    2    1         IF(J=2) THEN
 16    2    1            IF(I=2) THEN GO TO LEND;
 18    2    1         ELSE DO; I=I+1;
 20    2    2            GO TO SEVEN; END;
 22    2    2         ELSE DO; J=J+1; GO TO EIGHT; END;
 26    2    1         END; K=K+1; GO TO NINE; END;
 27    2    2
 31    2    1    LEND:END;
 32    1         END;
```

This shows a program similar to that shown in Figure 10-19, but written in compiler language PL/C, a "dialect" (variant) of PL/I. This program was actually run on a computer.

FIGURE 10-19 (c)

This shows a program similar to that shown in Figure 10-19, but written in compiler language **PL/C**, a "dialect" (variant) of PL/I. This program was actually run on a computer.

Exercise 10–49. A computer is used to total bills. For each item it reads the cost and adds a 4% sales tax if the item is taxable. Make a control chart and a FORTRAN program for this process. There should be a decision diamond corresponding to the question "Is this item taxable?"

EXAMPLE   A computer is programmed to assist instruction in arithmetic. Part of the program is shown in Figure 10–20. In this case the instruction READ means that the computer is to accept teletyped answers from the student, *I*1, the input (student's answer) for problem 1; *I*2, the student's answer for problem 2, and so forth. The instruction PRINT means that the information is to appear on the terminal for the student to read.

Figure 10–21 shows a flow chart for the program of Figure 10–20 on Computer Assisted Instruction. If the student types in the correct answer *I*1 for the first problem, then the computer takes him immediately to problem 10, which is of the next level of difficulty, requiring carrying a ten from the units addition. If the student types the wrong answer for problem 1, then the computer tries him on another of the same type. If he also misses that one, he is taken back to the first problem. If he gets the second one right, then he is given a third problem of the same type. If he misses problem 3, he may have guessed the answer to problem 2 without understanding what he was doing, so he is sent back to problem 1. If he is successful with problem 3, probably his mistake on the first problem was a careless error only, and he goes forward to problem 10.

```
1 PRINT,   21
             +43
             ‾‾‾
  READ, I1
  IF (I1.EQ.64) GO TO 10
  PRINT,   61
             +23
             ‾‾‾
  READ, I2
  IF (I2.NE.84) GO TO 1
  PRINT,   34
             +54
             ‾‾‾
  READ, I3
  IF (I3.NE.88) GO TO 1
10 PRINT,   39
             + 3
             ‾‾‾
            (continued . . .)
```

FIGURE 10–20. FRAGMENT OF A CAI FORTRAN PROGRAM.

The program for a CAI application needs to be more complicated than this example. For instance, a student who needs more practice with a skill would usually be given similar exercises, not identical ones to do over again. Also, a STOP is needed in case a student fails after many tries to use a technique. In such a case provision must be made for taking him even further back in the sequence, or for giving him personal tutoring.

Exercise 10–50. Describe your method of studying for a test over several skills. Suppose your first attempt to apply the first skill is successful. Do you ordinarily go on to the next skill? What if your first attempt is unsuccessful?

Exercise 10–51. Compare the process of Figure 10–21 with practicing

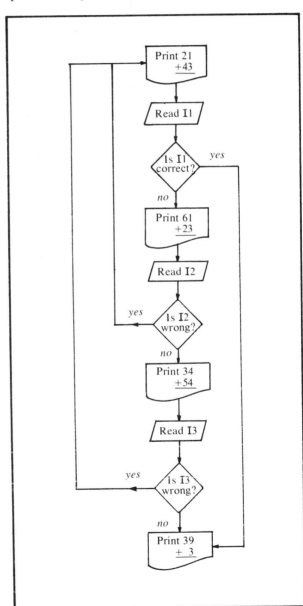

FIGURE 10–21. FLOW CHART FOR CAI PROGRAM IN FIGURE 10–20.

to acquire physical skills, such as driving a golf ball or putting, playing exercises of increasing difficulty on a musical instrument, or learning dance steps.

REVIEW OF CHAPTER 10

1. Discuss ways in which computers affect you now. Where do these applications need improvement?

2. How would you like to see computers used in the future?

3. Do you find it degrading to be known by a number, such as a student identification number? How can computers invade one's privacy?

4. Draw and identify typical sign shapes for flow charts.

5. What is the purpose of the arrows from a decision step?

6. Give an example of a process whose flow chart has a loop.

7. Draw a flow chart for the card game of War: The game calls for two players, who start by dividing a shuffled bridge deck of cards evenly between them. They match cards one pair at a time, high card taking the lower card. In case the cards turned over by the two players match, each turns over another card. When the tie is finally broken the high card takes all the tie cards that have accumulated. The game ends when one player has all the cards.

8. Study the flow chart in Figure 10–17. Try to write the corresponding FORTRAN program. Check by referring to Figure 10–18.

9. In Figure 10–18 find examples of transfer statements, of an arithmetic transfer statement, of a real variable, and of a real constant.

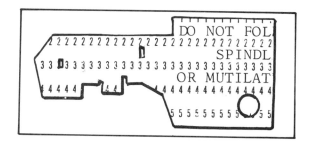

CHAPTER 10 AS A KEY TO MATHEMATICS

Perhaps you have had the experience of explaining something to another person and suddenly realizing that you understand it better yourself because of your explanation. In constructing flow charts and FORTRAN programs you have been explaining solutions of certain mathematical problems to a computer, in the process analyzing the mathematics so closely that you may find you understand it better than ever before.

Familiarity with some of the basic vocabulary may make available and interesting to you articles in the press about computing. This updated information can equip you to benefit from new applications of computers as they are invented. It can also help you protect yourself from some dangers present in the rapid computer expansion, invasion of privacy, and the threat to human feelings of individual identity and worth.

ANSWERS TO SELECTED EXERCISES

CHAPTER 1

1–1, p. 4. Diana. Cathy. Alice.

1–3, p. 5. Diana ranked Bob 3 and Cal 2, so reject Bob for her. Then link Bob to his next (second) choice, Alice.

1–7, p. 8. In both, Alice is linked to Bob and Cathy is linked to Don. In the boys' optimal assignment Al dates his first choice, Betty, but in the girls' optimal assignment he dates only his second choice, Diana. However, the girls' assignment dates Diana to her favorite, Al, and Betty to her favorite, Cal. Cal dates his favorite, Diana, under the boys' assignment, but dates his second choice, Betty, under the girls' assignment.

1–11, p. 11. In the final assignment, Duffy and Norris go to the Marines, Montcalm to the Army, James to the Navy. (partial answer)

1–13, p. 19. Spend all morning hours on science.

1–15, p. 20. To balance the classes exactly, assign 54 well-prepared students and all the disadvantaged students to Hope, 103 W to each of the other teachers. To take fullest advantage of the allowable discrepancy of 5 students, assign all the disadvantaged students to Hope, and either 101 W to Faith, 57 W to Hope, 102 W to Charity, or 100 W to F, 56 W to H, 104 W to C.

1–18, p. 24. There are chains from x to every other point.

1–21, p. 25. An arc from z to x, or one from z to y will make the graph strongly connected.

1–28, p. 33. Arcs are necessary, since, for instance, Estelle mentioned Jackie, but Jackie did not mention Estelle.

1–32, p. 33. Try to assign each girl to a patrol with girls she mentioned Resolve conflicts (crowded patrols) in favor of unpopular girls, rejecting girls who were chosen frequently.

1–38, p. 34. You might assign a high value, say 3, to pairing Norma and Pat to their choices, since neither was chosen by anyone. Then you might assign a value of 2 for good pairings for Leona, Sophie, and Barbara, who were chosen only once each. Then if you tried moving Leona from one patrol to another and it gave her one of her choices, you would add 2 points. If it took her away from her other two choices, you would also subtract 4 points, 2 for each one, making a net loss of 2 points for Leona in the change. Other pairings could have value 1. Notice that Exercise 1–36 suggested changing patrol assignments, with the possibility of placing a substantial value on assigning unpopular girls to leaders who chose them. The features on which you decide to place a high value are up to you. The whole question of popularity might be secondary to training and interest in nature study, water safety, first aid.

CHAPTER 2

2–3, p. 39. It was 5 o'clock 22 hours ago and 2410 hours ago.

2–6, p. 40. The 36-to-48 second pattern is exactly like the 0-to-12 second pattern, so the stencil could be laid along the 36-to-48 section, axis line matching, and the curve drawn through the stencil. To continue the graph from 25.7 seconds, place the stencil on the 24-to-36 second section, then at 12 second intervals.

2–10, p. 42. $m = 2$, $m = 5$, $m = 10$, $m = 25$, $m = 50$.

2–13, p. 42. $0, 2, 4, 6, 8, 10, -2, -4, -6, -8, -10, \ldots$, all even integers.

2–17, p. 43. If $a \equiv 0 \pmod{n}$, then $a = 0 + kn = k(qm) = kq(m)$, a multiple of m. Then $a \equiv 0 \pmod{m}$.

2–19, p. 45. $105 \equiv 0$

$6 \equiv 1$

$-3 \equiv 2$

$13 \equiv 3$

$24 \equiv 4$

2–21, p. 45. $\bar{0}$ contains $-5670, 0, 45, 75, 105$

$\bar{1}$ contains $16, 21, 11, 1111, 76, 71$

$\bar{2}$ contains $127, -8, -123, -3, 92, 12, 32$

$\bar{3}$ contains $-17, -7, 33, -27$

$\bar{4}$ contains $99, 29, 19, 4, 34, 14$

2–25, p. 47.

+	0	1	−1
0	0	1	−1
1	1	−1	0
−1	−1	0	1

×	0	1	−1
0	0	0	0
1	0	1	−1
−1	0	−1	1

2–28, p. 47.

×	0	1	2	3	−3	−2	−1
0	0	0	0	0	0	0	0
1	0	1	2	3	−3	−2	−1
2	0	2	−3	−1	1	3	−2
3	0	3	−1	2	−2	1	−3
−3	0	−3	1	−2	2	−1	3
−2	0	−2	3	1	−1	−3	2
−1	0	−1	−2	−3	3	2	1

2–30, p. 47. $2^2 = 4$. $2^3 = 8 \equiv -1$, so $2^4 = 2(2^3) \equiv -2$.
Then $2^4(3 + 12 + 15 + 1) \equiv (-2)(3 + 3 - 3 + 1) \equiv -8 \equiv 1$.
(partial answer)

2–32, p. 48. Your change in coins for a 32¢ item if you pay in bills is 68¢.
($32¢ + 68¢ \equiv 0$ modulo 1.00)

2–34, p. 48. $1, -1, 2, -2, 3, -3, 6, -6, 9, -9, 18, -18$. Division by zero is not defined, so zero is not a divisor. Zero is a multiple of 18, however, for $0 = 18(0)$.

2–40, p. 49. $32141 = 3 \cdot 10^4 + 2 \cdot 10^3 + 1 \cdot 10^2 + 4 \cdot 10 + 1$
$$\equiv 3 \cdot 0 + 2 \cdot 0 + 1 \cdot 0 + 4 \cdot 0 + 1 \equiv 1, \text{ not } 0.$$

2–45, p. 51. The second sum from the left is wrong. To find why it seemed to check modulo 9, find by how many it is off: correct sum $-2052 = $?

2–49, p. 52. 4.

2–52, p. 54. $6 = 4 \cdot 18 + (-1)66$.

2–53, p. 55. I expected 8 to attend, counting myself, but the actual number was 3 or 7, probably 7.

2–57, p. 57. Choose $x = 2 + 3 \cdot 5 = 17$, which will allow 14 repetitions of the shorter message.

2–58, p. 57. How many repetitions will $x = 2 + 25$ blocks of time allow
you? $\dfrac{27 \cdot 30 - 160}{25} = 26$. Too few.
If you buy $x = 2 + 2 \cdot 25 = 52$ blocks of 30 seconds, you will have
$\dfrac{52 \cdot 30 - 160}{25} = 56$. Still too few.
The smallest number of blocks that will allow 100 or more repetitions is $2 + 3 \cdot 5 + 3 \cdot 25$, or 92 blocks of time.

2–69, p. 59. $4x \equiv 2$ (mod 5), or $-x \equiv 2$, so $x \equiv -2$ (mod 5). The solutions are $x \equiv 3$ or 8 (mod 10).

2–74, p. 59. $(2, 6) = 2$ does not divide -3. No solutions.

2–77, p. 59. $x \equiv 0$ (mod 4).

2–80, p. 59. $8x \equiv 1$ (mod 11). By experiment, $8(4) = 32 \equiv -1$, so $8(-4) \equiv +1$. $x \equiv -4 \equiv 7$ (mod 11).

CHAPTER 3

3–2, p. 68. The first and third sentences are statements, even though they have truth value False for some of the people in the space. The second sentence is not decidable.

3–4, p. 69. (c) and (d) are not decidable. (f) can be made decidable by inserting the word "entirely." (g) would need further description as to size of holes or porosity to be decidable. (h) is not about objects in U.

3–7, p. 71. x has brains and beauty.

P	Q		$P \wedge Q$
T	T	brains and beauty	T
T	F	brains but no beauty	F
F	T	no brains but beauty	F
F	F	no brains, no beauty	F

3–11, p. 71. Here the logical connective can be used to give the points that lie in the intersection of Main and Park, as $M(x) \wedge P(x)$.

3–15, p. 73. M = male, G = graduate

M	G	
T	T	male graduate
T	F	male undergraduate
F	T	female graduate
F	F	female undergraduate

3–20, p. 75.

P	Q	$\sim P$	$\sim Q$	$(\sim P) \vee (\sim Q)$
T	T	F	F	F
T	F	F	T	T
F	T	T	F	T
F	F	T	T	T

SHADED PORTION REP-
RESENTS $(\sim P) \vee$
$(\sim Q)$. BIRDS THAT
SWIM, UNSHADED.

3–21, p. 75. $Q(x)$ is True for Buick, Ford, and DeSoto. (partial answer)

3–28, p. 77.   If x likes people and dogs, then x has lots of friends.

3–36, p. 78.   The biconditional is True for the first four pairs.

3–41, p. 79.   The vehicles in P (that is, vehicles x for which $P(x)$ is True) or in Q or both make both parts of the biconditional False, and hence make the biconditional itself True. The locomotive and the transportation-only car make both parts of the biconditional True.

3–48, p. 81. $[L(x) \wedge D(x)] \Rightarrow L(x)$. That required book is long. A second conclusion is: That required book is dull.

3–55, p. 83.

P	$\sim P$	$P \wedge (\sim P)$	$\sim[P \wedge (\sim P)]$
T	F	F	T
F	T	F	T

A statement has only one truth value, not both.

3–62, p. 86.

P	Q	$P \wedge Q$	$\sim(P \wedge Q)$	$(\sim P) \vee (\sim Q)$	equivalence
T	T	T	F	F	T
T	F	F	T	T	T
F	T	F	T	T	T
F	F	F	T	T	T

3–72, p. 88.   Of course not!

3–82, p. 89.   If snoodge, then smudgem.

3–92, p. 91.   Suppose Jack can eat no fat. Then together with 3, this means that the hypothesis of 1 is True (a tautology), so by the Law of Detachment, the platter and the pan are clean, contradicting 2.

3–99, p. 94. Reduce each premise to a tautological implication:
(1) $B \Rightarrow \sim P$ (in box implies not pencil)
(2) $C \Rightarrow \sim S$ (cigar implies not sugar-plum)
(3) $\sim B \Rightarrow C$ (not in box implies cigar)
The contrapositive of (2) is $S \Rightarrow \sim C$; the contrapositive of (3) is $\sim C \Rightarrow B$. We have the syllogism $S \Rightarrow \sim C \Rightarrow B \Rightarrow \sim P$, so that $S \Rightarrow \sim P$, the conclusion.

3–106, p. 95. Valid, despite fictions.

CHAPTER 4

4–2, p. 102. 1

4–10, p. 104. $u \lor (l \lor g') = u \lor g' = g'$, and
$(u \lor l) \lor g' = g' \lor g' = g'$.
$u \land (l \land g') = u \land l = 0$, and
$(u \land l) \land g' = 0 \land g' = 0$.

4–15, p. 106. 4, 6, 8, 12, 16, 18, 20, 24, 28, 30, 32, 36

$\begin{cases} 4, 8, 12, 16, 20, 24, 28, 32, 36 \\ 6, 12, 18, 24, 30, 36 \end{cases}$

4–23, p. 109. $\bar{C} \cup T = \{1, 2, 3, 5\}$. $\bar{C} \cup \bar{T} = \{1, 2, 3, 4, 5\}$.

4–29, p. 111. $X \cup Z = \{x \mid x$ gets up to 70 or between 20 and 30$\}$, which is equivalent to X. $W \cap X = W$.
$W \cap Y = \{x \mid x$ gets up to 35 and over 10, that is, between 10 and 35, not including 10$\}$. (partial answer)

4–31, p. 114. The middle and right circuits are equal.

4–44, p. 119. A circuit made of circuit a and circuit b connected in parallel is open if and only if both circuit a and circuit b are open.

4–46, p. 119. An element of the space that is not in the intersection of two sets must fail to be in one set or the other or both.

4–49, p. 121. One member of S, 1, is in T, but the other is not, so $S \nsubseteq T$. (partial answer)

CHAPTER 5

5–3, p. 130. $23 + 34 + 8 = 65$.

5–9, p. 130. $258(\frac{1}{4}) + 49(\frac{1}{2}) = 64\frac{1}{2} + 24\frac{1}{2} = 89$.

5–13, p. 131. $3(5) + 8(6) = 15 + 48 = 63$.

5–19, p. 136. $AB = \begin{pmatrix} 0 & 2 \\ \frac{3}{2} & -\frac{7}{4} \end{pmatrix}$. $BA = \begin{pmatrix} -2 & 1 \\ \frac{5}{2} & \frac{1}{4} \end{pmatrix}$. (partial answer)

5–26, p. 136. $(A + B)C = \begin{pmatrix} 2\frac{1}{2} & 8 \\ -1 & 2 \end{pmatrix} \cdot \begin{pmatrix} -2 & -3 \\ 0 & 1 \end{pmatrix} = \begin{pmatrix} -5 & \frac{1}{2} \\ 2 & 5 \end{pmatrix}$, which

should equal the sum of AC from Exercise 5–23 and BC from Exercise 5–24.

5–34, p. 137. $\left[\begin{pmatrix} 0 & 1 \\ 2 & 6 \end{pmatrix} \begin{pmatrix} 1 & 0 \\ 7 & 5 \end{pmatrix} \right] \begin{pmatrix} 2 \\ 3 \end{pmatrix} = \begin{pmatrix} 7 & 5 \\ 44 & 30 \end{pmatrix} \begin{pmatrix} 2 \\ 3 \end{pmatrix} = \begin{pmatrix} 29 \\ 178 \end{pmatrix}$. (partial answer)

5–37, p. 139. $7.41. $11.28. Adequate reader.

5–43, p. 141. 2 lb 2 oz $= 2(16) + 2$ oz $= 34$ oz $= 4$ oz $+ 30$ oz
$= 4$ oz $+ 7\frac{1}{2}$ (4 oz).

$\mathbf{w} = \begin{pmatrix} 1 \\ 8 \end{pmatrix}$. (partial answer)

5–46, p. 142. Abe 58. If he gets the full 20 points possible for homework and class participation, he will have 78. Cal 79$\frac{1}{2}$ for a possible 99$\frac{1}{2}$. (partial answer)

5–49, p. 143. Use an 8 by 8 matrix. (hint)

5–55, p. 146.

	1	2	3	4	5	6
d	1	1	1	0	0	0
e	1	0	0	1	0	1
f	0	1	0	0	1	1
g	0	0	1	1	1	0

5–64, p. 150. $C^{-1} = \begin{pmatrix} \frac{6}{13} & \frac{-1}{13} \\ \frac{1}{13} & \frac{2}{13} \end{pmatrix}$

5–68, p. 151. $\mathbf{r} = \begin{pmatrix} \text{hours} \\ \text{dollars} \end{pmatrix} = \begin{pmatrix} -\frac{1}{4} & 1 \\ \frac{3}{8} & -1 \end{pmatrix} \mathbf{f}$

Decrease boating $\frac{1}{4}$ unit, increase singing 1 unit.
Increase boating 1 unit, decrease singing 4 units.

CHAPTER 6

6–6, p. 159. c

6–31, p. 163. i (A.P.) $j = j + (j - i) = 2j - i$. (partial answer)

6–34, p. 163. Every pair of sets has a unique union and a unique intersection, provided we use all the subsets of the space. Even if two sets do not overlap, their intersection is the unique "null" set (having no members), $\varnothing$.

6–46, p. 169.

×	1	2	4	3
1	1	2	4	3
2	2	4	3	1
4	4	3	1	2
3	3	1	2	4

$2^1 \equiv 2,\ 2^2 \equiv 4,\ 2^3 \equiv 3,\ 2^4 \equiv 1$

6–55, p. 171. s, h, v, d, a

6–57, p. 171. $RH = A.\ HR = D.\ RD = H.$ (partial answer)

6–63, p. 175. $P^{-1} = \begin{pmatrix} 2 & 3 & 1 & 5 & 4 \\ 1 & 2 & 3 & 4 & 5 \end{pmatrix}$, which is usually rearranged to show the first row in numerical order, as $P^{-1} = \begin{pmatrix} 1 & 2 & 3 & 4 & 5 \\ 3 & 1 & 2 & 5 & 4 \end{pmatrix}$.

$$Q^{-1} = \begin{pmatrix} 2 & 1 & 3 & 5 & 4 \\ 1 & 2 & 3 & 4 & 5 \end{pmatrix} = \begin{pmatrix} 1 & 2 & 3 & 4 & 5 \\ 2 & 1 & 3 & 5 & 4 \end{pmatrix}.$$

$$P^{-1} \circ P = Q^{-1} \circ Q = Q \circ Q^{-1} = \begin{pmatrix} 1 & 2 & 3 & 4 & 5 \\ 1 & 2 & 3 & 4 & 5 \end{pmatrix}.$$

6–70, p. 178. Every multiple of 10 is an integer (in I). Each two multiples of 10 have exactly one sum, which is also a multiple of 10: $10i + 10j = 10(i + j)$. The identity is 0. The additive inverse of $10i$ is $10(-i)$. Then $\langle 10 \rangle$ is a group, and a subgroup of I. Similarly, $\langle 5 \rangle$ is a group. Every multiple of 10 is a multiple of 5: $10i = 5(2i)$, so $\langle 10 \rangle$ is a subgroup of $\langle 5 \rangle$.
Do not punch units digits.
Nickel.
Dime.

6–79, p. 182. If $n = km$, m has no inverse, for if there were an element m^{-1} for which $mm^{-1} \equiv 1$, we would have
$km = n \equiv 0$
$(km)m^{-1} = k(mm^{-1}) = k(1) = k \equiv 0$, but $k \not\equiv 0 \pmod{n}$.

CHAPTER 7

7–4, p. 188. $D(1, 1), G(-1, -1), J(1, -2), M(1, 2), P(-2, 2), S(-3, -2),$ $V(4, -2)$. (partial answer)

7–11, p. 191. Yellow $y = 2\frac{1}{2}$ and $y = 3$
Purple $y = \frac{1}{2}$ and $y = 1$
Green $y = -1\frac{1}{2}$ and $y = -\frac{1}{2}$
Black (upper stripe) $y = -2$ and $y = -1\frac{1}{2}$ (partial answer)

7–15, p. 194.

$$F \begin{cases} y - 1 = 0 & \text{from } x = -3 \text{ to } x = -2 \\ y = 0 & \text{from } x = -3 \text{ to } x = -2 \\ x + 3 = 0 & \text{from } y = -1 \text{ to } y = 1 \end{cases}$$

(partial answer)

7–21, p. 198.   (partial answer)

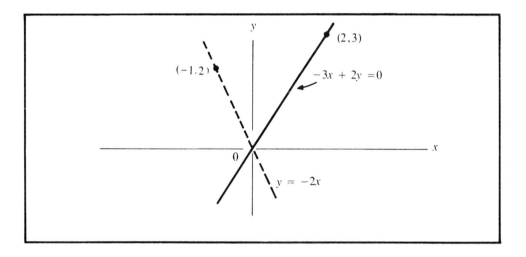

7–25, p. 199.   Necessary rate is $280/3\frac{1}{2} = 80$ mph. This line, $d = 80t$, passes through the point $(3\frac{1}{2}, 280)$. (partial answer)

7–30, p. 203.   *OB*: $y = x$. *EA*: $y = x + 2$. *DC*: $y = x - 3$. $b = 0, 2, -3$, respectively.

7–34, p. 205.   Should resemble the key drawn for the "Key" section at the end of Chapter 7.

7–38, p. 207. $y = -2x + 8 = 4x - 12$. $6x = 20$. $x = \frac{10}{3}$. $y = \frac{4}{3}$.

7–42, p. 208.   20 hours, 1000 miles.

7–47, p. 211. X: $(2, -2)$. U: $(9, 0)$
$XU = \sqrt{(9 - 2)^2 + (0 - (-2))^2} = \sqrt{7^2 + 2^2} = \sqrt{53}.$
$7^2 = 49$, $8^2 = 64$. $(7.2)^2 = 51.84$, $(7.3)^2 = 53.29$.
$\sqrt{53}$ is between 7.2 and 7.3. (partial answer)

7–50, p. 212. $(x - 1)^2 + (y + 5)^2 = -10 + 1 + 25 = 16$. Center: $(1, -5)$, radius 4.

CHAPTER 8

8–11, p. 225. $\frac{1}{3} + \frac{1}{3} = \frac{2}{3} > \frac{1}{2}$, polyhedron made of 3-gons (triangles), 3 at each vertex. Tetrahedron.

$\frac{1}{3} + \frac{1}{5}$ gives the icosahedron, $\frac{1}{5} + \frac{1}{3}$ the dodecahedron. (partial answer)

CHAPTER 9

9–1, p. 233. You would lose the benefits of method N for those students who need it. The school system might get a reputation for not trying new techniques, might find it increasingly difficult to get support for experimentation, might lose teachers who are innovative.

9–3, p. 233. Sentencing an innocent person *versus* risk to society of acquitting a guilty person.

9–6, p. 233. I.Q. may be influenced by cultural background, health at time of taking test, cooperativeness of subject. It is used for want of a better test to estimate "native" intelligence, not learned knowledge.

9–17, p. 237. No. Previous records show the relative frequency of fatalities on weekends, during commuters' hours, and so on.

9–22, p. 243. Yes, significant at the 0.05 level of significance.

9–27, p. 249. α. β.

9–29, p. 251. Pair students with similar attitudes. If you estimate that about $\frac{1}{4}$ of the population has a favorable attitude toward N, $\frac{1}{4}$ against N, $\frac{1}{2}$ apathetic, then the sample should contain about the same proportion of pairs of each attitude: $\frac{1}{4}$ for N, $\frac{1}{4}$ against N, $\frac{1}{2}$ apathetic.

9–34, p. 253. The decision changes at 25%, when the average cost is $\$1 \times (25\%$ of 20$) = \$5$. (partial answer)

9–38, p. 254. $[(90\%)(90\%)][(90\%)(90\%)][(90\%)(90\%)][(90\%)(90\%)]$
$= (81\%)(81\%)(81\%)(81\%)$
$= (65.61\%)(65.61\%) = $ almost 43.05%

CHAPTER 10

10–21, p. 280. $a = H.$ yes to $c = d.$ $b = W.$ yes to e or $f = X$, no to the other (f or e) $= Y$.

10–25, p. 283. The general pattern of Figure 10–13 works well for this flow chart: Start. Assign 1 to N. Read E, H. Assign $10 to P. Is H.GT.5? If *yes*, add $2(H − 5) to P, and Print E, P. If *no*, Print E, P. Is N.EQ.15? If *no*, add 1 to N and return to "Read E, H." If *yes*, Stop.

10–29, p. 288. Connector C, when $k = 1$. Conflict chart.

10–46, p. 302. Let JS stand for Student, N for number of correct answers on the main test, KB for number (0 or 1) of correct bonus question answers, LG for grade.

```
   I = 1
 5 READ, JS, N, KB
   LG = 10 * N
   IF (KB.GT.0) GO TO 10
   GO TO 20
10 LG = LG + 15
20 PRINT, JS, LG
   IF (I.EQ.1355) GO TO 30
   I = I + 1
   GO TO 5
30 STOP
   END
```

10–48, p. 302. Let *JP* stand for the price of each ticket, 0, 2, or 4. Let *KT* stand for the total of admissions.

```
   KT = 0
   I = 1
 5 READ, JP
   L = JP - 2
   IF (L) 10, 20, 30
10 KT = KT + 0
   GO TO 40
20 KT = KT + 2
   GO TO 40
30 KT = KT + 4
40 IF (I.EQ.970) GO TO 50
   I = I + 1
   GO TO 5
50 STOP
   END
```

INDEX

Page on which definition appears is distinguished by **boldface** type.

SYMBOLS